Trust and Disinformation

Michael M. Resch
Michael Herrmann • Andreas Kaminski
Maike Stelzer • Jörn Wiengarn
Editors

Trust and Disinformation

Springer

Editors
Michael M. Resch
University of Stuttgart
High Performance Computing Center
Stuttgart, Baden-Württemberg, Germany

Andreas Kaminski
Philosophy Department
TU Darmstadt
Darmstadt, Germany

Jörn Wiengarn
Philosophy Department
TU Darmstadt
Darmstadt, Germany

Michael Herrmann
Tübingen Forum for Science and Humanities
University of Tübingen
Tübingen, Germany

Maike Stelzer
KRITIS
TU Darmstadt
Darmstadt, Germany

ISBN 978-3-031-96789-4 ISBN 978-3-031-96790-0 (eBook)
https://doi.org/10.1007/978-3-031-96790-0

This Springer imprint is published by the registered company Springer Nature Switzerland AG
The registered company address is: Gewerbestrasse 11, 6330 Cham, Switzerland

Contents

Explanations and Counter-Measures

About the Authors

Brian Ball is an Associate Professor in Philosophy, and Research Lead for Artificial Intelligence and Information Ethics, at Northeastern University London. In addition to his work in the philosophy of AI and (mis)information, he has published extensively in the theory of knowledge, philosophical psychology, and the philosophy of language.

Casey Doyle is Lecturer in Philosophy at Binghamton University in New York, USA. He works at the intersections of philosophy of mind and epistemology, with a focus on self-knowledge. His additional research interests include philosophy of technology and certain areas of history. He received his PhD from the University of Pittsburgh.

Sacha Ferrari is a doctoral researcher at the Center for Logic and Philosophy of Science of the Institute of Philosophy at KU Leuven. Sacha Ferrari received a Master's degree in Physics and a Master's degree in Philosophy from the Free University of Brussels. His PhD aims to analyze non-institutionalized scientific practices (such as neuro hacking, self-medicine, and DIY biology) in the light of the concept of individualism and autonomy. His particular area of interest focuses on social epistemology, computational epistemology, and STS.

Michael Herrmann is a researcher and head of the Tübingen Forum for Science and Humanities at the University of Tübingen. Previously he was affiliated to the High Performance Computing Center at the University of Stuttgart (HLRS). His research lies in the philosophy of science and addresses philosophical questions with respect to computer-intensive approximative methods. He is particularly interested in epistemological foundations of computer simulations and machine learning. His current work is about methodological changes and different styles of reasoning between classical statistics and machine learning and the role of value-ladenness in (numerical) algorithms.

Andreas Kaminski is a professor of philosophy of science and technology at the Technical University of Darmstadt. His research areas include social epistemology (especially the philosophy of trust and testimony), technical epistemology (the role of technology in science like computer simulation and machine learning), and politics of technology (here especially modeling for policy). Andreas Kaminski is co-editor of the *Yearbook of Philosophy of Technology*. He is a senior scientist at the High-Performance Computing Center of the University of Stuttgart (HLRS), where he established and led a department for the philosophy of computer-based sciences. Previously, he was a visiting professor at RWTH Aachen University.

Alexandros Koliousis is an Associate Professor in Computer Science at Northeastern University London. He works on the design and implementation of scalable AI systems, models, and algorithms with applications to social network analysis, neural network verification and testing, image processing, natural language processing, and network management.

Marcel Erik Lemmer is a research associate at Karlshochschule International University and a PhD candidate at Goethe University. His research focuses on the constitution and contestation of state and collective narratives, examining their impact and discourse on both international and national scales.

Lei Niu is a PhD student in philosophy at the University of Cologne. He works on social and political epistemology. He is particularly interested in issues concerning complementary methods to rational persuasion in combating fake news, conspiracy theories, and people's irrationality.

Amil Mohanan is an Assistant Professor in Data Science at Northeastern University London. He received his PhD in Geography at the University College London in 2015 researching computer networks and governmentality. Amil has previously held research roles at The Open University and Loughborough University where he studied Financialization and the Geographies of Higher Education.

Viet Anh Nguyen Duc is a research associate at the Institute of Philosophy at TU Darmstadt. The focus of his research is on the ironic self and world relations as they manifest in public. He received his PhD from the TU Darmstadt.

Mike Peacey is a Senior Lecturer in Economics at the University of Bristol. He is a microeconomic theorist with an interest in modeling a diverse range of real-world situations. For example, higher education markets, regulation of betting exchanges, and social media and misinformation.

Andrija Šoć is a research associate at the Institute of Philosophy in Belgrade. His current main area of focus is connecting the areas of political theory, political science, and empirical research, with the focus on philosophical analysis of conceptual underpinnings of these different fields. In 2021 and 2023, he taught MA courses at the Charles University in Prague. He has also worked on different topics in history of philosophy, his publications focusing mostly on various aspects of Kant's philosophy. His recent research focuses, among other topics, on examining participatory models of democracy, and interdisciplinary approach to the problem of trust.

Jörn Wiengarn is a research associate at the Institute of Philosophy at TU Darmstadt. His research is centered on social-epistemological aspect of science and technology, especially focusing on the dynamics of trust and mistrust within these contexts. Jörn Wiengarn is co-editor of the *Yearbook of Philosophy of Technology*. He earned his PhD from the University of Cologne.

Oliver Zöllner (M.A., U of Bochum, 1993; Dr. phil., U of Bochum, 1996) is a professor of media research, media sociology, digital ethics, and international communication at Stuttgart Media University (Hochschule der Medien) in Germany where he is co-director of the Institute for Digital Ethics. As an honorary professor, he also teaches at Heinrich Heine University Düsseldorf. From 1997 to 2004, Zöllner served as director of audience research at Germany's international broadcaster, *Deutsche Welle*. His research interests focus on questions of digitalization, digital transformation, and related aspects of reflexive media literacy.

Trust, Testimony and Technology: What Makes It So Hard to Combat Disinformation?

Michael Herrmann, Andreas Kaminski, and Jörn Wiengarn

Contents

Abstract In this introduction, we seek to highlight the central role of trust in combating disinformation. Currently, four main narratives dominate the discourse on disinformation: an epistemic narrative, a technological narrative, a normative narrative, and a social narrative. Each of these offers a distinct explanation of the problem and suggests a different approach to addressing it. Yet, as we argue, the success of all these approaches ultimately depends on whom potential recipients of disinformation decide to trust. Trust, we suggest, is an often overlooked but crucial dimension in the effort to mitigate disinformation. In a further step, we connect these four narratives to corresponding notions of trust developed in the philosophical literature. We also briefly examine how recent advances in image-generative AI – including the rise of deepfakes – have expanded the domain of testimony, thereby heightening the importance of trust in acquiring information. We then outline both the philosophical and practical difficulties involved in defining disinformation. And finally, we will develop an argument showing that trust is always embedded within holistic networks of trust and distrust spanning multiple sources and institutions.

All authors contributed equally to this work.

M. Herrmann
Tübingen Forum for Science and Humanities, University of Tübingen, Tübingen, Germany
e-mail: michael.herrmann@tfw.uni-tuebingen.de

A. Kaminski · J. Wiengarn (✉)
Philosophy Department, TU Darmstadt, Darmstadt, Germany
e-mail: andreas.kaminski@tu-darmstadt.de; joern.wiengarn@tu-darmstadt.de

M. M. Resch et al. (eds.), *Trust and Disinformation*,
https://doi.org/10.1007/978-3-031-96790-0_1

The fact that disinformation constitutes a major problem in today's world hardly needs elaboration. Disinformation presents challenges on multiple levels: it can lead individuals to hold false beliefs, sometimes with harmful consequences for both themselves and others. It can also shape collective discourses, diverting attention from pressing issues. Moreover, disinformation can generate higher-order confusion in the public sphere about what can and cannot be believed.

Notably, disinformation poses a problem specifically for democratic societies, impacting their institutions and public discourses. This stems from the fact that, unlike autocracies, democracies are committed to openness and diversity in the public sphere. Certain forms of control and intervention employed by autocratic states to combat disinformation are therefore neither viable nor desirable from a democratic perspective. At the same time, democracies depend on reliable information for public debate and collective decision-making. Consequently, democracies – unlike autocracies – must publicly negotiate and debate a normative tension between the value of reliable information and the freedom of expression, a tension that also carries legal and democratic-theoretical implications.

Beyond these normative questions, though, one might initially assume that, *on a purely practical level,* combating disinformation would be straightforward. After all, it seems to involve nothing more than removing false information or correcting it with accurate facts. However, experience shows that addressing disinformation is not that simple and poses a more persistent problem. But why is this so? What makes it so hard to combat disinformation?

To explore this question, we begin with adopting a simple heuristic model of the socio-epistemological situation in which disinformation becomes a problem. This model assumes a basic testimonial situation where a recipient of information A (for example, readers of news in the digital realm) encounter an information provider B (such as news outlets, social media accounts that share information, etc.) who assert an alleged fact p. This configuration is the basic backdrop for the phenomenon of disinformation. Put in the language of testimony (Fig. 1):

Now, there are basically two further possibilities, each with roughly two options. A can either trust B or A can distrust B. And: the communicated information p can either be true or false. Thus, we have (Figs. 2, 3):

At a first pass, we can now specify the conditions that give rise to problematic disinformation, leading to false beliefs: It is precisely when conditions (2.1) and (3.2) hold at the same time, i.e. *when A trust B that p even though p is false.* This simple, preliminary model will be used in the following as a heuristic to outline different approaches to addressing disinformation. Yet, as we will see in the course of

Fig. 1 Testimony

Fig. 2 Testimony and trust

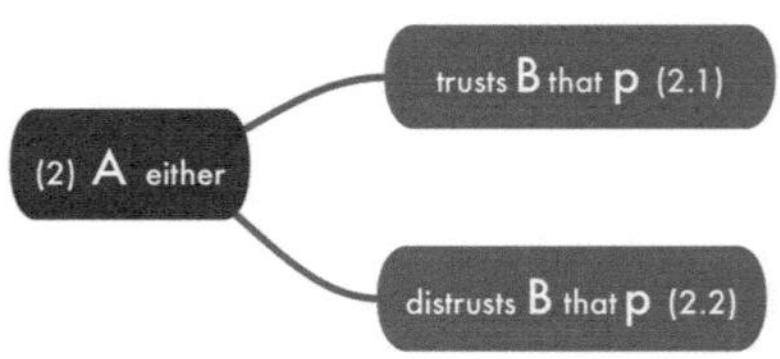

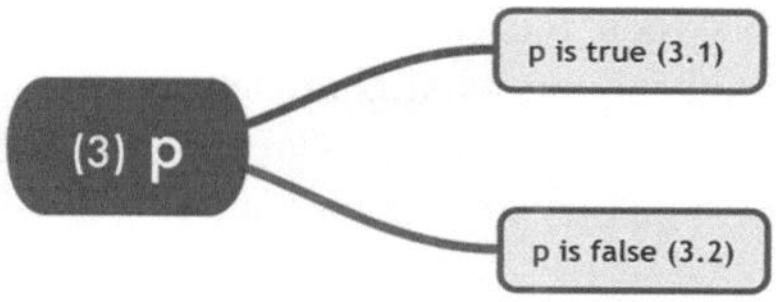

Fig. 3 Testimony and truth

our argumentation, the model will quickly prove to be insufficient and will need to be gradually refined and enriched.

1 Disinformation: Four Narratives, Four Sources, Four Solution Approaches

What we take to be effective strategies for addressing disinformation depends on what we identify as its root cause. Currently, four narratives about the rise of disinformation prevail, each identifying a different source. By attributing different causes to the rise and success of disinformation, they reveal different threats and lead to substantially different solution approaches.

1.1 The Epistemic Narrative

The first narrative starts directly from the definition of the problem itself: disinformation is, at its core, deceptive and misleading information.[1] Consequently, individuals adopt false beliefs or feel that, in the face of diverse and contradictory claims, they must refrain from making any judgment – even when accurate information is available. Within this narrative, the threat of disinformation is described in purely epistemological terms: It poses a danger to the process of acquiring knowledge. Under the traditional conception of knowledge as "justified true belief", disinformation undermines each of its components: It fosters false beliefs or obstructs the formation of beliefs altogether, and it compromises the justification condition of knowledge. The justification of a belief depends on the reliability of the process through which it is formed. Yet, the proliferation of fake news or seemingly realistic but artificially created photographs or videos renders these belief-forming processes unreliable.

From this perspective, what else can be done to combat disinformation other than countering it with accurate information? Within the framework of the epistemic narrative, replacing disinformation by correct information appears to be the obvious solution. This, however, raises a critical question: how can one reliably distinguish between correct and deceptive information? The epistemic approach offers a simple answer: through knowledge. The difference between what is true and what is false can and must be recognized epistemically, typically on the basis of evidence. Even when deceptive information is presented as accurate, it is possible to differentiate

[1] Or, depending on the definition, no information at all. Cf. Floridi, 2019, 80–82.

between the two. The available evidence provides the means to discern what is truthful and what is misleading.

We have thus introduced the idea that evidence serves as a foundation. The amount and quality of evidence we possess determines the degree of trust we are justified to have in the assertion that p. On this view, evidence is conceived as independent of trust. Conversely, the rationality of trust depends on the evidence it is based on (Fig. 4).

From here, we arrive at a first transformation of our above notation. What A must do is gather evidence that B is trustworthy or that p is credible (Fig. 5):

Placed within the framework of our basic model above, we can say that the core mechanism of the epistemic approach is that A should base their trust on evidence indicating that p is actually true, or, conversely, that A should distrust when they have evidence that p is false. Evidence thus plays an intermediary role here, aiming to align A's trust (2.1) with the truth of p (3.1) or their distrust (2.2) with the falsehood of p (3.2) (Fig. 6).

The limitations of this approach, though, can be well illustrated with the following example. Since at least 2008, a conspiracy theory has circulated in the United States claiming that Barack Obama was not born in the U.S. but in Kenya. According to the proponents of this conspiracy theory, this would disqualify Obama as a natural-born citizen and thus exclude him from holding the office of the U.S. President (however, it is important to note that even this is legally disputed). This racist conspiracy theory led a significant portion of the population to believe that Obama had indeed been born in Kenya. In an effort to debunk this myth, Obama and his team decided to publish Obama's birth certificate, an act that, due to its official nature, can be regarded as the presentation of *authenticated evidence*. However, as is well known, this evidence failed to silence public doubts. As an April

Fig. 4 A first idea: Evidence is the foundation of the trustworthiness of information

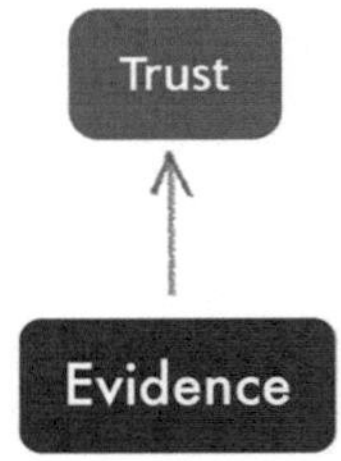

Fig. 5 The epistemic solution

Fig. 6 The aim of combating disinformation

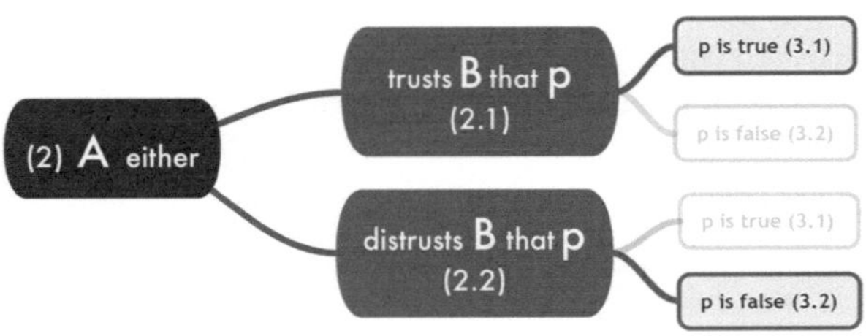

2011 Gallup poll revealed, 24% of respondents still believed Obama was probably or definitely born outside the United States. Even after Obama subsequently released the long-form version of his birth certificate a few days later, a following Gallup poll in May showed that public doubts decreased but did not vanish entirely. This persistence was – at least in part – certainly fueled by Trump's malicious tweets, in which he claimed that the birth certificate was a forgery, and pointed to alleged phantom numbers on the document (Fig. 7).

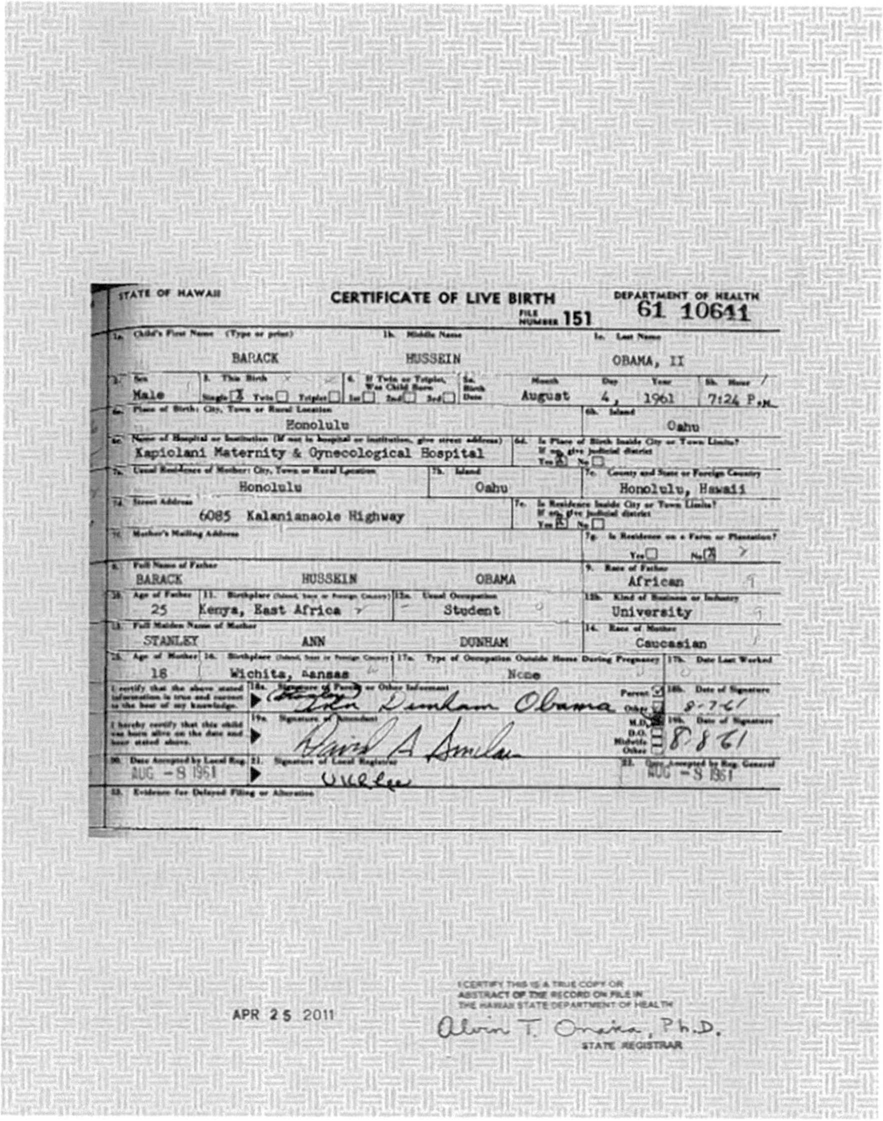

STATE OF HAWAII
CERTIFICATE OF LIVE BIRTH
DEPARTMENT OF HEALTH
FILE NUMBER 151 61 10641

BARACK HUSSEIN OBAMA, II
Male Single August 4, 1961 7:24 P.M.
Honolulu Oahu
Kapiolani Maternity & Gynecological Hospital
Honolulu Oahu Honolulu, Hawaii
6085 Kalanianaole Highway
BARACK HUSSEIN OBAMA African
25 Kenya, East Africa Student University
STANLEY ANN DUNHAM Caucasian
18 Wichita, Kansas None
8-7-61
8-8-61
AUG - 8 1961
AUG - 8 1961

I CERTIFY THIS IS A TRUE COPY OR ABSTRACT OF THE RECORD ON FILE IN THE HAWAII STATE DEPARTMENT OF HEALTH
APR 25 2011
Alvin T. Onaka, Ph.D.
STATE REGISTRAR

Fig. 7 Long form of Obama's birth certificate https://obamawhitehouse.archives.gov/sites/default/1ies/rss_Viewer/birth-certificate-long-form.pdf Accessed 23 Aug 2024

This example offers an important theoretical insight. It demonstrates a limitation of the epistemic approach, since it rests on a crucial but hidden assumption. Namely: Evidence is usually presented *by others* and, as in the case of Obama's birth certificate, *authenticated by them* as evidence. Therefore, evidence can*not be recognized* as evidence without *being acknowledged as such,* and *acknowledging* the evidence, in turn, presupposes *trust* in the source presenting it (Fig. 8).[2]

Consequently, the main institutions tasked with finding, analyzing, and evaluating evidence – science and journalism – can only fulfill this role under one condition: they must be perceived as *trustworthy* sources. And this is why the epistemic approach to the problem of disinformation relies on trust as a tacit requirement.

We started with the idea that evidence provides a basis for justified trust in information. However, we now see that evidence itself requires trust in the source presenting it in order to be recognized as evidence (Fig. 9).

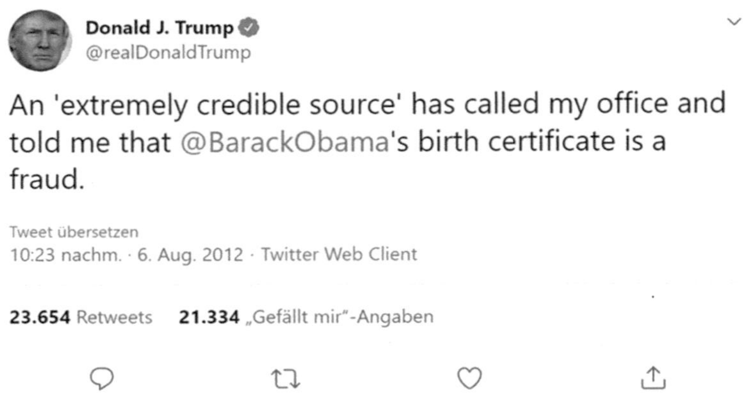

Fig. 8 Trump's malicious tweet. https://twitter.com/realDonaldTrump/status/232572505238433794 Accessed 23 Aug 2024

Fig. 9 The supporting role of trust

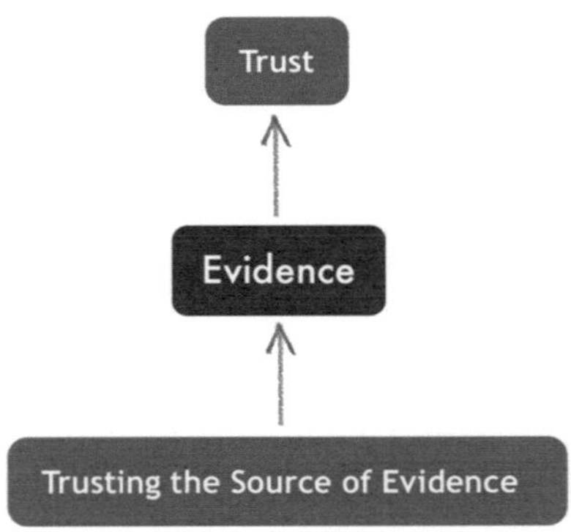

[2] Bernd Lahno has formulated this argument for the philosophy of trust in opposition to epistemic approaches (Lahno 2002, 138).

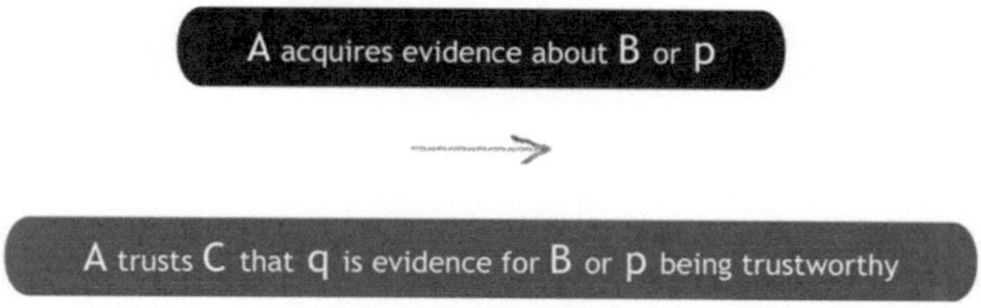

Fig. 10 Unavoidable circles of trust

This changes our second formulation substantially. We need to broaden our perspective and turn the foundation from its head (evidence) to its feet (trust). In most cases, A will only gain evidence for the trustworthiness or untrustworthiness of B or p by trusting C (another person, organization, or institution), who provides A with evidence of B's trustworthiness (or lack thereof) (Fig. 10):

1.2 The Technological Narrative

In addition to the epistemic approach, but closely related to it, a technological narrative has emerged. According to this narrative, the current sources of disinformation are rooted primarily in new technologies that enable and facilitate it. To understand how novel and far-reaching today's technical possibilities are, one only needs to consider how extensive the toolbox has become that individuals, organizations, and states now have at their disposal in order to create, disseminate, and even measure the success of disinformation:

- *Imitations:* Social bots that pretend to be real people; Generative Adversarial Networks (GANs) whose synthetically generated faces, voices, or actions (deep fakes) resemble real photos or videos.
- *Decoys:* Incentives (such as games or free apps) that enable the collection of data (including through geofencing) to create user profiles (e.g., for mass personalization); decoys work by diverting users' attention.
- *Camouflages:* Social networks where content selection is driven by user profiling, data analytics and adaptive methods, while the selection process itself remains opaque.
- *Potemkin networks:* Social media content generated by bots and adaptively curated for users, which create the illusion of societal majorities that appear underrepresented outside the network. This fosters the belief that traditional media are biased and should be dismissed as 'mainstream'.

If technology plays a central role in the spread of disinformation, it stands to reason to combat it technologically as well. Numerous attempts are currently being undertaken in this regard. These efforts are closely connected to epistemic approaches, as they mostly rely on technical means of detecting and then either flagging or automatically removing disinformation. However, the rapid advancement of technology highlights its double-edged nature – where technology can both fuel and counter disinformation. This dynamic tends to result in an arms race between the purveyors of disinformation and those who are fighting against it. A clear example of this is deepfake detection, where the same technology used to create deepfakes is also leveraged to detect them.

This, in turn, reframes the problem of disinformation and how it should be addressed. Instead of C serving as the source of evidence, technology T now adopts this role by detecting whether B or p are deceptive (Fig. 11):

Again, expressed within the framework of our basic model, we can say that technology plays the role of encouraging A to distrust B that p (2.2) when there are reasons to believe that p is likely false (3.2) (for example by flagging content); technological measures my even prevent A from encountering the information from B at all (suspending 1) if p is false (3.2) (in the case of deleting false content). Thus, in the case of flagging, technology aids in aligning A's decision to trust with the truth of B's statement, while also aligning A's distrust with the falsehood of *p* (Fig. 12):

Technical detection of disinformation or fake news can include, for example, analyzing the propagation patterns of messages and the interconnectivity of actors in social networks in order to identify networks of bots. Bots not only offer a cost-effective way to drastically increase the volume of messages, but they also have an immense social-psychological advantage. Numerous studies in social psychology have shown that people tend to align their beliefs (or at least the expression of their beliefs) with the represented majority of opinions (cf. Asch, 1951, 1956 and Franzen & Mader, 2023). In this way, *fake* majorities, created by bots pretending to be real people, can evolve into *genuine* majorities of real people who adapt their beliefs to these perceived majorities. Additionally, such fake majorities are likely to be

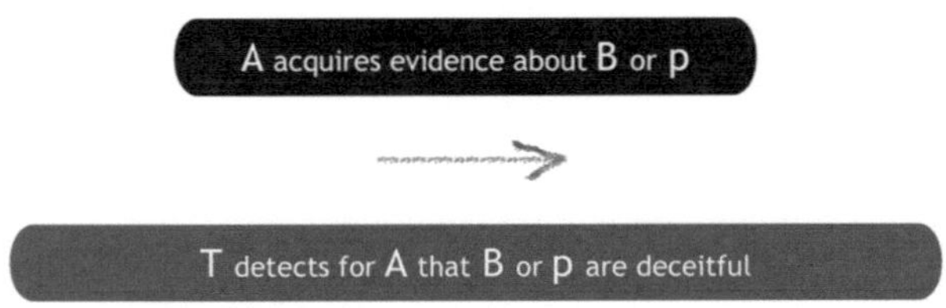

Fig. 11 The technical solution

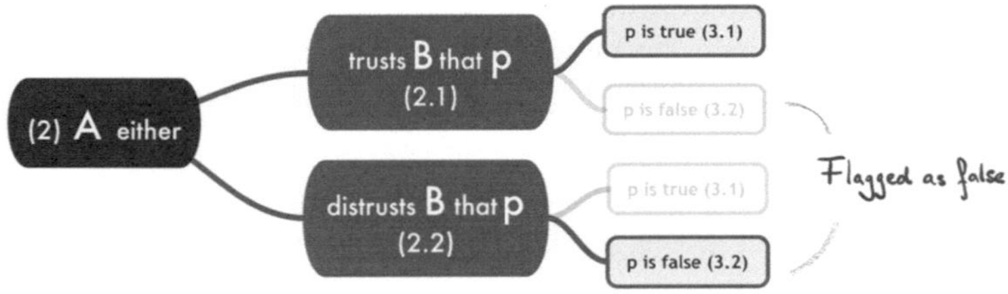

Fig. 12 Automatic detection, flagging or removal of false statements

perveived as not adequately represented in traditional journalistic media. The resulting discrepancy between the perceived majority opinion on social media and the perspectives presented in traditional media outlets can, in turn, foster mistrust in the latter.

An alternative technical approach focuses on evaluating the trustworthiness of the information itself. Such methods include an analysis of textual elements (such as language, style, headlines, vocabulary, expressiveness, the use of quotes or images, etc.) as possible indicators to assess the trustworthiness of putative news. Using machine learning techniques, models are trained to detect fake news, with the reliability of specific indicators for recognizing fake news becoming apparent during the training process. One of the key limitations of this approach leads directly back to the previously discussed limitation of the epistemic approach – this technical solution also *requires trust*: trust in the tools and thus in the organizations and institutions that develop and deploy them. Therefore, the core problem of the technical approach can be reduced to the epistemic problem. While it might seem that an additional dimension arises – the question of technological reliability – this concern is not unique to technology. It applies equally to personal testimony as a source of evidence, as introduced in the epistemic approach: testimonial sources, too, must be reliable (Fig. 13).

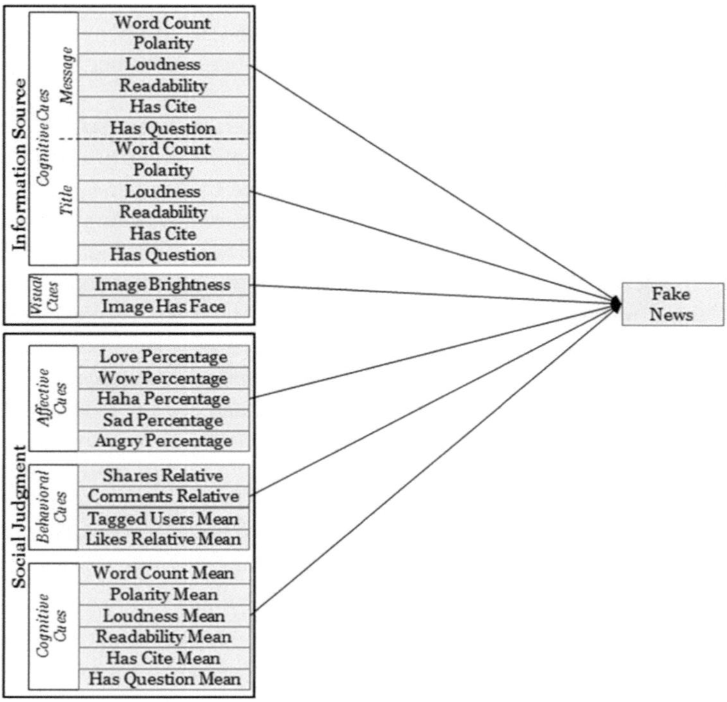

Fig. 13 Exploratory research model to identify fake news by Janz and Risius 2017, 5

1.3 The Normative Narrative

A rather different approach focuses on the normativity of communication. Even when speakers are not explicitly aware of it, communication is deeply and intricately structured by norms. This was expressed, for example, by Grice with his idea of conversational maxims (Grice 1975). To illustrate their relevance in communication processes, consider the following example: if you ask where Stephanie is right now and your roommate responds by saying that there's a red car in front of the house, you usually won't be puzzled – even though the literal sense of the response appears unrelated to the question; after all, the question was about Stephanie's location, not about a red car. Grice's point, however, was that such conversational exchanges reveal the influence of communicative norms. Your roommate's statement is perceived as an adequate answer because the norm of communicating *relevant* information allows for the evident inference that Stephanie probably drives a red car and is therefore at home. A notable example of the skillful exploitation of such norms can be seen in Mike Pence's speech at the Republican National Convention in 2020. The then Vice President spoke about an officer of the Department of Homeland Security who was shot during violent left-wing protests against police brutality in Oakland, California. According to Grice's communication norms, Pence implied that the officer was killed by politically left-wing protesters and that his death was directly connected to the protests. Pence repeatedly stressed the importance of "law and order" in his speech, further reinforcing this impression. In reality, however, the arrested suspect was a member of the far-right Boogaloo group.[3]

Importantly, communication is structured differently on social networks. Through their socio-technical design, the meaning and weight of communicative norms change. A theoretical insight into this is offered by Nguyen (2021). His central idea is based on a philosophical analysis of games. According to Nguyen, in games, the values of actions are determined by the ways in which one can win. Similarly, the design of communicative spaces leads to a gamification of social networks. Central to this are scoring mechanisms that quantitatively evaluate users' contributions – for instance, on Twitter (now X), through likes, retweets, and followers. Nguyen argues that these scoring systems shift communicative values. In other words, people adapt their communication style and content to align with these scoring mechanisms and to communicate successfully as defined by them. In particular, messages that are "loud" and attention-grabbing tend to score highly, incentivizing corresponding sensationalist communication.

In a nutshell, the problem is that the new communicative spaces either lack certain norms or that they alter communicate values through the way they are designed (Fig. 14):

[3] Cf. for the full transcript of Pence's speech: https://www.nytimes.com/2020/08/26/us/politics/mike-pence-rnc-speech.html. Accessed 13 September 2024.

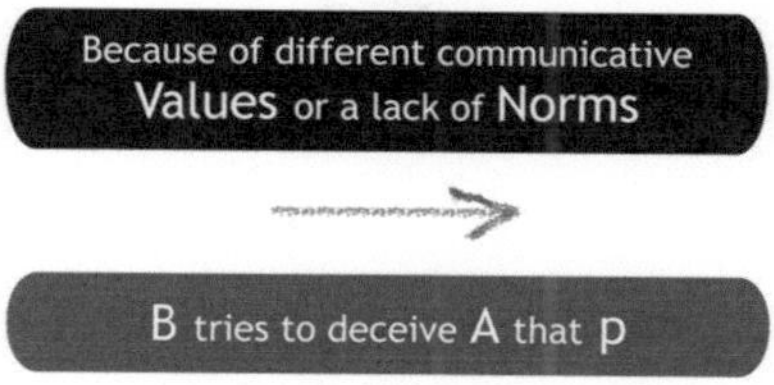

Fig. 14 The normative approach

The fact that people communicate differently in spaces that are normatively structured differently is well exemplified by the court cases in the United States following the 2020 presidential election. For example, Rudy Giuliani claimed on television that there had been "massive fraud".[4] Similarly, Sidney Powell stated that there was a widespread conspiracy brought about by the "massive influence of communist money." In court, however, Trump's representatives seemed to tell a different story. During a hearing before a judge in Pennsylvania, Giuliani admitted: "This is not a fraud case." And when the judge asked Trump's attorney Jonathan S. Goldstein whether he was alleging election fraud, Goldstein responded: "Your honor, accusing people of fraud is a pretty big step (…). And it is rare that I call somebody a liar, and I am not calling the Board of the [Democratic National Committee] or anybody else involved in this a liar. Everybody is coming to this with good faith." The judge then pressed him further: "Are you claiming that there is any fraud in connection with these 592 disputed ballots?" To this, Goldstein ultimately responded: "To my knowledge at present, no."

What can account for this striking discrepancy between the statements? Apparently: The role of communicative norms. Yet, if these norms are so powerful, it seems reasonable to use them to combat disinformation. The basic idea is simple: we need to redesign communicative socio-technical spaces, shaped by different scoring mechanisms. Pennycook et al. (2021) showed that even small nudges have a significant effect on how often false news is shared. When users who had previously shared false news were subsequently asked whether they believed a message was true or false, this small intervention affected how often they shared false news with others within the following 24 hours. The explanation provided by the authors is that the question shifted users' attention, specifically, to the truth of the message. An alternative, complementary explanation is that the intervention directed attention to the value of truthful communication, thereby enhancing users' sense of responsibility for adhering to this communicative norm.

According to this narrative, the solution lies in normatively restructuring the communication space (Fig. 15):

[4] These and the other quotes are taken from the article "Donald Trump and His Lawyers Are Making Sweeping Allegations of Voter Fraud in Public. In Court, They Say No Such Thing." https://time.com/5914377/donald-trump-no-evidence-fraud. Accessed 13 September 2024.

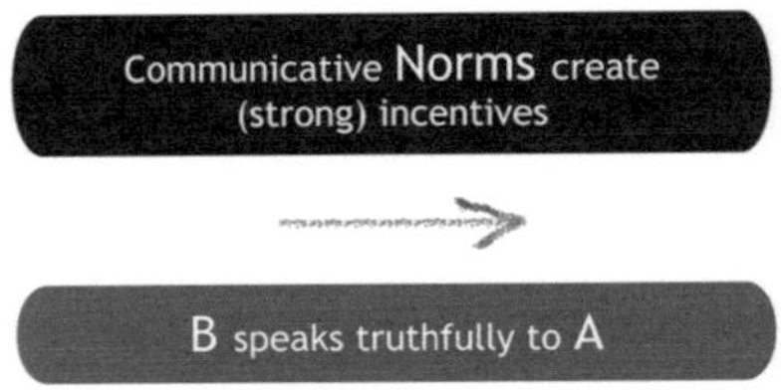

Fig. 15 The normative solution

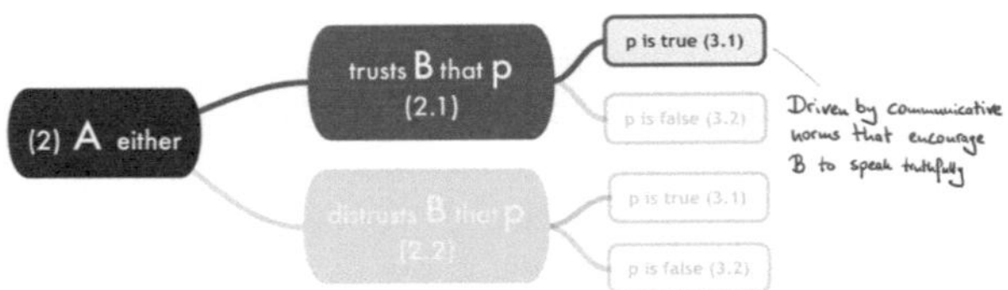

Fig. 16 Goals of the normative approach

If we wanted to apply this basic idea to our model above, we could say that this approach seeks to prevent actors from making false claims in the first place. In effect, this strengthens the connection between acts of communication (1) and the truth of what is communicated (3.1). Moreover, when *A* is aware of these strong communicative norms, it provides them with a reason to trust in the testimony of *B* (Fig. 16):

However, the core idea of altering the significance and weight of communicative norms through the design of communicative spaces raises a critical question: who has the authority to make these decisions and how can these decisions be justified? Should social network operators be responsible for them? And if so, how could any decision by them be justified, given the central role that social networks now play in democratic discourse.

For addressing this question, two considerations seem important:

1. Any explicit structuring is, quite understandably, bound to invite criticism and to arouse suspicions of manipulation.
2. Nevertheless, it must be noted that social networks are already engaging in this kind of structuring; the operators (owner, designers) of social networks are continuously making such decisions. It is only because they are not particularly visible, that they often go unchallenged.

In this sense, rejecting the normative structuring of social media (in the sense of: politics should keep its hands off my posts) resembles the (naive) belief that 'politics should keep its hands off the food on my plate'. The selection of food available in supermarkets is already heavily pre-structured, even if this pre-structuring is not particularly salient. Those who oppose 'politics' to 'control what's on my plate' overlook the fact that other hands already maintain quite a firm grip on it.

A possible solution could be to make the normative structuring of communication spaces itself a topic within these communication spaces, allowing for a co-evolution of communication and its communication space. However, this proposal encounters at least two major difficulties: first, given the way communication is currently structured online, it is unlikely that any objective discussion about it can be expected; and second, the owners of these platforms will likely resist such discussions if they threaten their business models.

1.4 The Social Narrative

The social narrative attributes disinformation to social factors, such as polarization, characterized by conspiracy thinking, filter bubbles, echo chambers or other social pathologies. The main concern is that the informational environment of individuals becomes so isolated that they are no longer confronted with differing or opposing world views and that alternative sources of information are discredited (cf. Nguyen, 2020). For example, in the case of conspiracy theories as sources of disinformation, experiences of epistemic injustice can create a form of seemingly rational mistrust toward epistemic authorities.

According to this narrative, democratic societies are increasingly organizing themselves along polarized, antagonistic political-identity groups. This fosters frictionless self-affirmation of beliefs within each group and the discrediting of opposing perspectives. As strong identities become increasingly linked to specific politically charged belief systems about the world, socially adaptive and identity-protective forms of cognition begin to outweigh objective truth-orientation (cf. e.g., Klein, 2020, Talisse, 2019 and Königs, 2024, cf. also Doyle in this volume).

How such polarization manifests, for example, in terms of center versus periphery, power versus powerlessness, or granted versus denied recognition from the perspective of those involved, is illustrated in a study by Harambam & Aupers (2015). In this study, individuals who had disseminated conspiracy theories online were extensively interviewed about their biographies and motivations. A recurring theme in these interviews was the participants' sense of experienced epistemic injustice.

Here, the question is not whether this perceived injustice is objectively justified; rather, what is noteworthy is that the participants themselves cite epistemic injustice as part of their motivating experience:

- "I am no medical expert, I am no doctor, so it's not true, right? They ask me what my scientific background is, so I tell them about the anthroposophic studies I completed in Germany. But that's not scientific, so it means nothing. Lately I've been asking my wife: 'Should I also attend university and get a medical degree? More people would believe me.' But why is that? Who decides that? I find it ridiculous."

- "But why? I am also a human being and I have done my study of life, so why? I have my own feelings and emotions and experience so why? Because you've studied you know how it works, right? When you haven't studied you don't count in this society …"
- "It's all like: 'I have studied, I am a doctor, I know more than you, so I will enlighten you. You are a layman.' So already from moment A there's a hierarchy, and they just instruct you to have your baby vaccinated, because well, that's procedure. So I said, 'listen, I've done my own research and I have this and that consideration.' And the nurse at the clinic just sits there and does exactly what she's learned to do: just copy and paste."
- "I consider it strange that people think: 'Oh these white coated people, they know it all, so we follow, we surrender.' Because they don't know it all! As a human being I can decide 'how', I want to stay in charge and don't want to surrender to doctors like that! I would like to have conversations about how we are going to fix things, what the other possibilities are."[5]

The quotations clearly express a perceived lack of acknowledgment from the perspective of the interviewees, experienced as (epistemic and social) injustice. This makes responding to them particularly challenging. Engaging epistemically with the claims expressed in the quotations might backfire and just reinforce the impression of epistemic injustice. It may even lead to people feeling lectured, which, from their perspective, would repeat their experience of injustice. Alternatively, a normative engagement with the parties involved might only succeed if their epistemic claims are recognized. Yet this would simply be the opposite of combating disinformation and amount to reinforcing false beliefs. If we assume that deep structural causes (such as polarization) cannot be easily changed, the next best step may be to seek social acknowledgement for individuals who drift into conspiracy theories due to experiences of injustice. The idea behind this approach is that it is social relationships that lead A to have a high level of distrust in B as an epistemic authority. Consequently, the solution to this can only consist in transforming (e.g., healing) these social relationships (perceived violations). However, there is no rulebook for achieving such acknowledgement (Fig. 17).

Expressed again within the framework of our model, the main idea of this approach is primarily to prevent actors from developing distrust (2.2) toward information that is, in fact, accurate (3.1).

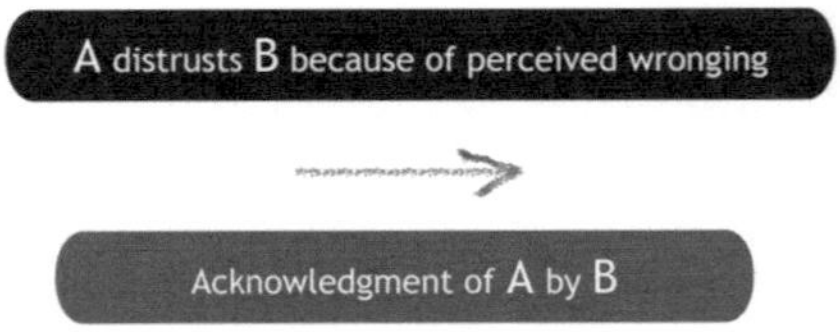

Fig. 17 The social approach and its solution

[5] All quotes are from Harambam & Aupers, 2015.

2 Different Narratives, Different Notions of Trust

The four different narratives correspond to three different notions of trust. Both, the epistemic and the technological narrative are based on an epistemic concept of trust, which is sometimes also referred to as the evidential view of trust. The normative narrative aligns with a normative concept of trust, but things are a bit more complicated here, as we shall soon see. Finally, the social narrative can be linked to a virtue-based conception of trust.

The differences between these conceptions of trust can be explained by considering how they conceive the relationship between trust and its reasons. Each of the four narratives about trust and disinformation are associated with distinct reasons on which trust is based (Fig. 18):

Let us begin with the idea of *knowledge* as a reason. On this view, A can possess knowledge about B that leads them to believe B's testimony that p. Such a knowledge-based reason may derive from a track record, an indicator, or a strategic consideration. For all these ways of acquiring knowledge, one can find corresponding approaches in the theoretical history of trust, for instance Hume's evidential view (Hume, 2007 [1748], 79 ff.), rational choice and game theory (Coleman, 1990, 99, Gambetta, 1988) or psychometrical approaches (Stern, 1903, Greuel et al., 1997 and Arntzen, 2007). Track records and indicators are often well-suited for *technical* implementations in computational models, particularly in agent-based and game-theoretical frameworks. Since these approaches interpret the disinformation problem as a problem of trust, and in turn conceptualize the trust problem as a knowledge problem, the same limitations that affect knowledge-based trust theories also apply to the epistemic and technological strategies for combating disinformation (Kaminski, 2017, 2021). As we have seen, knowledge about B (or about p directly) is typically provided by other parties C, D, or E as evidence for B's degree of trustworthiness. For this evidence to be accepted as valid, these other parties must themselves already be trusted.

Let us now turn to the third narrative and its link to the normative conception of trust. At first glance, this approach seems rather straightforward: The normative concept of trust posits that reasons are based on norms and values. This way of putting it, however, conceals, a potential source of misunderstanding. Even if one

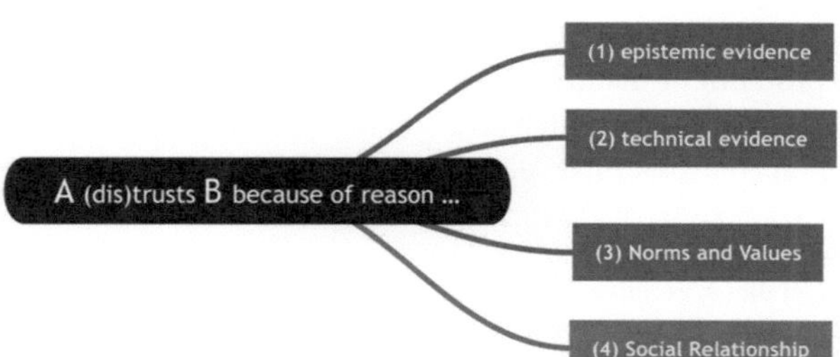

Fig. 18 Four narratives, four reasons to (Dis)trust

acknowledges the importance of normative structures for testimonial exchange, it is important to note that A may regard such structures either as normative or as epistemic reasons. To see what is meant by this, imagine that A believes B because B would face severe sanctions if it were proven that their statement that p were knowingly false. Thus, A knows that B knows that if B were to lie, B would face serious punishment. And A also knows (or can at least assume) that B wishes to avoid such punishment. Consequently, A has a specific reason to believe B that p. But this is *not* a normative reason. If there are any doubts about this, one only needs to go back two or three sentences. We said: A *knows* that B knows that if B were to lie, B would face serious punishment. And A also *knows* (or can at least assume) that B wants to avoid this punishment. Evidently, the role of the normative structure here is reduced to providing *epistemic* reasons for A as to why B can be believed. However, to make sense of a genuinely normative conception of trust, we need to understand what it means for A to have genuinely normative reasons for trusting B.

Here, the assurance view of testimony opens up a perspective on how this can be understood (Moran, 2006). However, it is questionable how this idea can help in developing an approach to combating disinformation. Everyone who wants to be believed will assure that they are telling the truth – and this specifically includes those who want to deceive others. B, who speaks truthfully, and C, who speaks deceptively, both assure A that what they say is accurate. The difference between the two may, in principle, be epistemically recognizable, but even if it is, that only means that the reason is no longer understood normatively.[6]

We can now turn to the fourth approach, which views social relationships between persons (and institutions) as having an inseparably epistemic and normative dimension. On this view, reasons are both epistemic and normative. The paradigm for this is found in virtue theory, where it has been shown that it is not possible to attribute a virtue to a person unless both (i.e., A and B) recognize and acknowledge corresponding behaviors as virtuous. Even though this approach promises to resolve theoretical paradoxes of the first two approaches, significant practical challenges remain. In highly differentiated societies, interactions are often too formal and the contexts too complex to understand how a person's virtuous behavior manifests itself. For example, we don't know many of the journalists who provide us with news and our understanding of what constitutes virtuous behavior for institutions or organizations is often vague and insufficient. Assessing virtuousness on one's own typically requires a certain level of closeness and familiarity with those we wish to evaluate. This means that virtue-based approaches cannot promise an easy fix to the problem of disinformation either.

In conclusion we can say: Each trust theory provides a different conception of how to arrive at reasons that allow distinguishing between trustworthy and

---

[6] This does not mean that normative considerations are out of play, but normative approaches cannot be integrated so easily. For an alternative to this, cf. Wiengarn, 2022 and Kaminski, 2017, 2021.

untrustworthy assertions. However, none can offer a direct solution to the problem of disinformation.

3 The Expansion of Testimony

Let us now return to our formal model of the basic testimonial situation. Another possible modification concerns p. Technological change leads to an expansion of the class of phenomena that appear as statements and thus as testimony. While this may initially seem surprising, it has become increasingly evident. In his influential essay "Getting Told and Being Believed" (2006), Richard Moran explains the difference between the evidential view (that is, epistemic theories of trust) and his own assurance view through a comparison between photography and testimony. A photograph, according to Moran, serves as evidence and speaks for itself. When we examine a photo, Moran argues, we are all in the same epistemic position: By looking at it we find evidence of what occurred at the time it was taken. Even the photographer might discover details they hadn't noticed before. Testimony, however, is different, according to Moran. It aims at the recognition of the speaker's intention; and it is only by recognizing this intention that we can say that one genuinely believes *a speaker* what they say. In this regard, the speaker and the listener do not occupy the same position.

Even back in 2006, the rigidity of this distinction could be questioned, and technological progress since then has only amplified such doubts. It has become increasingly common to present photos with disclaimers stating that they have not been altered or generated by AI models. The technical possibilities to alter or even generate pictures from scratch have led to a situation where they increasingly require support from additional testimony to fulfill their role as evidence of something. Thus, evidence does not speak for itself even in the case of photos, videos, or audio recordings but must be backed up by testimony and trust in testimonial sources. Under these circumstances, evidence requires assurance. Photos therefore tend to become a case of B telling us that p. Consequently, the scope of testimony, and therefore p, is expanding (Fig. 19).

Fig. 19 The expansion of p

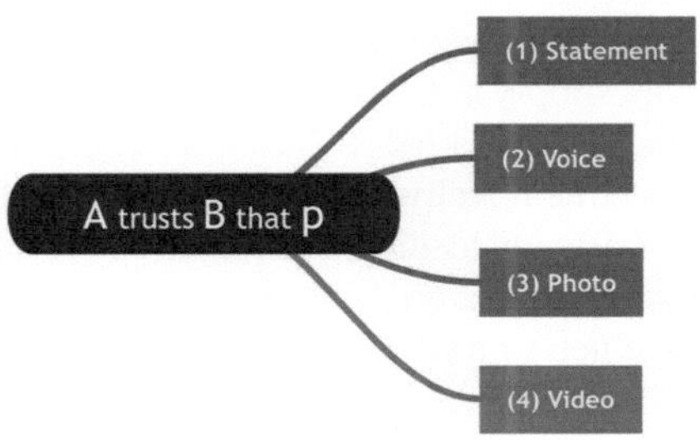

4 How to Define Disinformation and Its Sub-Types?

Another open question regarding how the trust relationship between A and B should be characterized in our above model also leads to the question of an appropriate definition of disinformation. Above, in the central model situation, we simplified by saying that the aspect in which A can trust B is that p is true. However, the scope of trust can actually be broadened to include that B also pursues certain *motives*.

Accordingly, there are different ways in which A's trust can be disappointed. One possibility is that p simply turns out to be false – even if the author B of the supposed information acted in good faith and to the best of their knowledge. In such a case B was simply mistaken. Another, arguably deeper, form of disappointment occurs when B reveals a genuine intent to deceive. The motives of those who produce information are also relevant to the question of an adequate definition of disinformation and its key subtypes, which will now be briefly addressed.

Disinformation is typically understood as a specific subset of misinformation: while misinformation refers to false information shared without intent, disinformation denotes false information spread deliberately. Noteworthy, this broad definition can capture a variety of quite distinct phenomena, including Fake News, rumors, deepfakes or other manipulated images, conspiracy theories, and, arguably, even hoaxes and satirical content. These forms are typically associated with social media as their primary platform of dissemination. However, the scope of disinformation extends further, encompassing deception in public speeches, misleading advertising, and other forms of strategic misinformation intended to manipulate or mislead.

A pressing issue with these various terms is the lack of standardized definitions, leading to some conceptual confusion. The debates surrounding the definition of "Fake News" serve as a symptomatic example of this. While there is general agreement that Fake News refers to information that is epistemically inaccurate and mimics the style of genuine news articles, the specifics of its definition remain contentious. A key point of debate is whether Fake News must be intentionally false, and if so, whether a specific motive on the part of its creators is necessary for its classification.[7]

While some have argued that Fake News must be distinguished from mere bad journalism or unintentional errors in legitimate news reporting by an intention to deceive or lie (cf. Gelfert, 2018, 105 f., Fallis & Mathiesen, 2019), others contend that this demand is too rigid. They propose instead that the producers of Fake News may simply not care for the truth – rendering it a form of "bullshit" (Mukerji, 2018). Axel Gelfert (2018, 2021), however, argues that this approach fails to address the systemic novelty of the phenomenon. He suggests that Fake News should be understood as misleading *by design*, where the phrase "by design" is meant to capture the "systemic features inherent in the design of the sources and channels through which fake news proliferates" (Gelfert, 2018, 109).

[7] For an overview, see Laster and Janius (2018; 2021). Tandoc Jr et al. (2018) also provide an overview and taxonomy, refering to a broader range of empirical disciplines.

Note that these conceptual debates are not confined to philosophical discourse but also play a significant role in the empirical study of Fake News and in the technical literature on Fake News detection, where scholars similarly grapple with how to conceptualize and define the term (cf. Baptista & Gradim, 2022, Tandoc Jr et al., 2018).

The debate becomes even more complex as some scholars reject the concept of Fake News altogether, arguing that it is too vague, unstable, and politicized to be useful (Coady, 2019 and Habgood-Coote, 2019).[8] Yet others, however, have suggested alternative methodological approaches to make sense of the term and its definition. In a recent paper, Gelfert (2021), for instance, proposes that Fake News can serve an exploratory function: even without a clear and stable definition, the concept can be useful for probing whether the phenomenon it refers to is coherent and novel. Furthermore, some have dismissed efforts to define Fake News solely by appealing to and explicating its ordinary meaning. Grundmann (2023), for example, advocates for a consumer-oriented definition, arguing that while it remains close to the meaning of "Fake News" in ordinary language, it also promises to be epistemically fruitful.

In light of these controversies and the quickly overwhelming complexity of the debate, we believe it is worth noting a few uncontroversial points. First, even in contexts where definitions are unclear, there is usually an uncontested semantic core that can be agreed upon – such as the idea that Fake News consists of information that mimics the format of news but is false. Second, it is certainly helpful to have an overview of the various, more specific definitional approaches available, perhaps in the form of a taxonomy – such as the one proposed by Jaster & Lanius (2018; 2021) for Fake News (also: Soc as well as Nguyen and Lemmer in this volume). Third, particularly when there is no well-justified or widely accepted definition of a term, it seems necessary, especially in empirical studies, to settle on a definition that best serves the research interests and the specific aspects one intends to investigate. In this context, it seems advisable to follow a pragmatic approach, explaining terms as precise technical definitions *tailored to the specific task at hand*. Consider the example of Fake News detection: in such contexts, it becomes evident that defining Fake News is not just a theoretical exercise, but one with tangible, practical implications. The chosen definition determines what qualifies as Fake News and, as a result, what may be banned, downranked, or flagged. In such practical contexts, it is essential that definitional decisions account for their potential consequences, rather than being made in isolation from them.

Lastly, note that there is an additional dimension to these conceptual challenges: beyond how a term should be defined, there is the question of how to *operationalize* it. This raises the issue of which specific instances of information should be classified as disinformation and thus *fall* under a given definition. These are not trivial matters. Take the case of Fake News again: even if we adopt a definition of it as information that mimics credible news while claiming falsehoods, there may still be

[8] For a reply to this ciriticism, cf. Bernecker et al. (2021, 4–6).

considerable ambiguity over whether a *specific* piece of content fits this definition. Uncertainty can arise around what a given piece of news actually claims, how it is intended to be understood, whether it is demonstrably false or simply uncertain, or whether one can read it as implying a falsehood. What makes these questions especially urgent is that disinformation often exploits exactly such ambiguities, making it difficult to classify it as such, and in doing so, potentially circumventing regulatory measures.[9]

5 Disinformation in the Context of Networks of Trust and Distrust

Finally, we revisit the above basic model of the testimonial situation once more to expand it from another perspective. For heuristic reasons, we assumed that A interacts with only *one information source B*. However, in reality, the landscape faced by actors navigating the information space is far more complex. Information consumers are not only confronted with a broad array of diverse testimonial actors – embedded within different institutional and organizational contexts that shape distinct incentive structures etc. Moreover, as we will explore in this final section, it seems even *necessary* for individuals orienting themselves in this landscape that they navigate within entire *networks of mutually supporting trust and distrust relationships with different actors*. In a way, such complex networks of trust and distrust serve as a (albeit contingent and malleable) transcendental background for information consumers to find orientation (cf. Baurmann & Cohnitz, 2021 and Arnold & Wiengarn, 2024). Gaining a clearer grasp of this broader socio-epistemological background could also help to get a better understanding of the conditions, mechanisms and detrimental effects of disinformation.

5.1 *Trust as an Irreducible Condition of Knowledge*

Quite generally, navigating the news world and the broader information landscape fundamentally depends on trust. We come to know facts about the world – from climate change to vaccine safety – only through trust in the sources and experts that present us with information. To gain knowledge about most kinds of information presented in the media, trust is indispensable. After all, how can a layperson conduct independent research to verify the reality of climate change or the efficacy of vaccines? It is not only impractical, but simply impossible, for individuals to reach such conclusions based solely on personal observation or one's own scientific reasoning. At some point, we must trust the available data and its interpretations provided by experts.

[9] Cf. also Wiengarn (forthcoming) on this topic.

In fact, in a highly specialized and technologically advanced society like ours, even experts rely on trust. Specialists in one field often depend on the established findings and technologies of other domains, as well as the work of their colleagues.

This dependence on trust also implies a significant limitation in the common advice to "do one's own research"[10] in response to disinformation. While a skeptical approach may appear prudent, the idea that one can *completely* suspend trust and independently verify all claims is a misconception. Research itself relies on networks of trust (Hardwig, 1991), whether in the reliability of data, the credibility of sources, or the peer-reviewed work of other scholars. Thus, "doing one's own research" is not an alternative to trusting – doing research is itself deeply embedded within a web of trust relationships – just like most of our knowledge. On a normative level, recognizing the irreducibility of trust relationships reveals why "do your own research" is a flawed advice. Instead, the focus should be on promoting rational and *responsible trust* in navigating the information landscape. However, defining what it means to trust rationally and responsibly is a complex challenge and an appealing research desideratum. One particular challenge is how to avoid circular reasoning in this regard, as we will elaborate now.

5.2 Networks of Trust and Distrust

Importantly, trust rarely exists in isolation. When we trust a piece of news, for instance, we are implicitly trusting a network of other agents and institutions. First, we trust the source of the news, assuming there is a reliable institutional framework ensuring the accuracy of the report. Often, media institutions are assumed to function as gatekeepers in this regard (cf. Mößner, 2018, Fallis and Mathiesen 2020 on the importance of this institutional dimension). However, such a trusting attitude seems itself to be built on other layers of trust. After all, how do we know these institutions exist and operate as expected? This knowledge cannot typically be gained through direct observation either but relies on further trust in additional sources that inform us about these institutions. As argued above, assessing the truth of a piece of information, p, by examining the evidence for p similarly presupposes an initial trust. One must already trust the source providing this evidence. And how in turn can this trust in the source itself be epistemically justified?

These brief sketches of argumentation underscore that trust in any piece of information is rarely, if ever, isolated. Instead, it appears embedded within holistic networks of mutually reinforcing trust relationships. This implies that our basic model outlined above must be expanded further (Fig. 20).

[10] On this phrase, cf.: Ballantyne et al., 2024

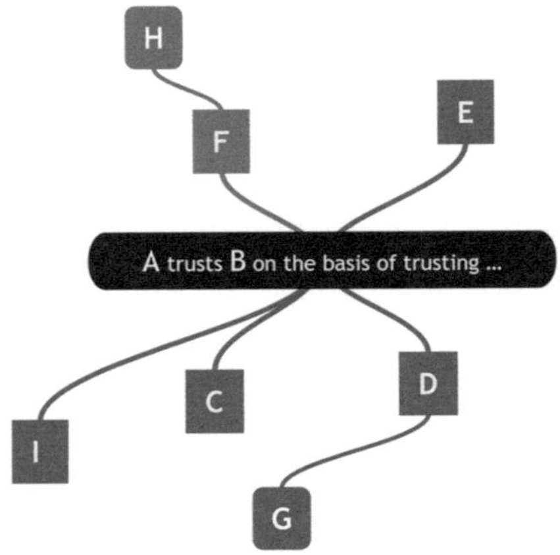

Fig. 20 Networks of trust

Note furthermore, that even distrust or skepticism, such as science denialism, ultimately rests on some form of trust in alternative sources. For instance, a climate change denier may base their rejection of scientific consensus on trust in *alternative* sources that claim to provide scientific testimony. As Wittgenstein famously argued, real doubt presupposes certainty: "If you tried to doubt everything you would not get as far as doubting anything. The game of doubting itself presupposes certainty" (Wittgenstein, 2015, §115). In a similar vein, one might say that every act of mistrust implicitly relies on trust in other information. This underscores that trust, even in its negative forms, is always part of a networked trust structure.

This perspective might offer a helpful framework for addressing key questions about disinformation. Thus, understanding the effects of disinformation as disruptions to people's trusting and distrusting attitudes provides insight into its broader impact. Disinformation doesn't just lead people to adopt false beliefs – it can confuse or reshape their trust networks, often fostering new forms of mistrust. Thus, by placing trust at the center of the analysis, we can better grasp the deeper structural consequences of disinformation. Bearing in mind the normative and affective dimension of trust, this perspective also promises to shed light on how trust dynamics contribute to phenomena like polarization and the formation of echo chambers (cf. Nguyen, 2020).

Overview

Part I: *Varieties of Disinformation* addresses the various forms of disinformation and how they can be conceptually distinguished. *Trust in Falsehoods: Definitions, Sources, Challenges* by **Andrija Šoć** critically examines the conceptual distinctions commonly made between misinformation, disinformation, and malinformation, arguing that these categories require a more nuanced taxonomy to account for all forms of falsehoods. The paper further explores how each form of deception – such as fake news, spin, rumors, and conspiracy theories – exploits different components of trust, whether cognitive, affective, motivational, or evaluative. Šoć argues that effective strategies for combating misinformation and disinformation must be tailored to the specific trust mechanisms they target. Additionally, the paper highlights emerging challenges posed by AI-generated content, which complicates the issue of trust in information even further.

When Homeopathy Cures Cancer: About Bullshit, Irony, and Disinformation in the Homeopathic Discourse by **Viet Anh Nguyen Duc** and **Marcel Erik Lemmer** explores different forms of disinformation in the context of homeopathy. The authors distinguish between two modalities of disinformation: a Strategic Modality, where disinformation is spread with the intent to deceive (often for personal or financial gain), and an Expressive Modality, where disinformation is used to articulate mistrust of established institutions, such as science or conventional medicine. Building on that, the paper develops a typology of five ideal-typical narrative carriers within the homeopathy discourse, each driven by different motives and employing distinct disinformation strategies. These include True Believers, Committed Persons, Bullshitters, Semi-Ironic Persons, and Entrepreneurs, with each type contributing to the spread of disinformation in different ways. Another variation of disinformation is explored in **Jörn Wiengarn**'s *How to Lie without Lying: Exploring Strategies of Assertion-Denial*. Wiengarn focuses on a specific type of strategy that is being used in the context of disinformation campaigns. This strategy allows speakers to disseminate falsehoods while evading responsibility by exploiting ambiguities in what qualifies as an assertive speech act. Wiengarn refers to this tactic as the strategy of assertion-denial, which allows speakers to retroactively reinterpret their statements as non-assertive. Common variants of this technique include claiming "It's just my opinion" or "That was meant sarcastically." Through several examples, Wiengarn illustrates how this strategy enables speakers to avoid being held accountable for spreading disinformation.

Part II: ***Describing the Post-truth Predicament*** is dedicated to diagnosing various aspects of our current epistemic malaise.

In **Casey Doyle**'s *Politization, Signaling and the Epistemic Landscape,* Doyle explains how politicization affects belief formation by shifting people's focus from truth-tracking cognition to socially adaptive cognition. In politicized contexts, where individual's beliefs become markers of group identity, individuals usually find themselves in an incentive structure that favors social alignment and political signaling over evidence-based inquiry. Doyle refers to examples like mask-wearing during the COVID-19 pandemic to show how this dynamic leads to polarization and undermines epistemic trust in experts and institutions.

In *Fake News and Conspiracy Narratives in the Context of Regimes of 'Posttruth': An Analysis from the Perspective of Nihilism* **Oliver Zöllner** explores how fake news and conspiracy theories are manifestations of what he calls a nihilistic attitude prevalent in posttruth societies. He argues that disinformation thrives in an environment no longer oriented towards objectivity, where reality is manipulated for political or ideological ends. Zöllner connects this phenomenon to the concept of kitsch – exaggerated forms of representation that imitate reality but strip it of its complexity and truth. He also draws on Jean Baudrillard's notion of hyperreality, where media and images construct a perception of reality detached from actual facts. Zöllner further emphasizes the role of social media algorithms and big tech companies in amplifying this trend.

Part III: ***Explanations and Counter-Measures*** shows, firstly, how social network analysis can explain the persistence of disinformation within a community, and secondly, adresses different perspectives on countering disinformation. In their paper *Ignorance in Social Networks: Discounting Delays and Shape Matters* **Brian Ball, Alexandros Koliousis, Amil Mohanan** and **Mike Peacey** investigate the impact of social network structure on belief formation dynamics by simulating scenarios in which disinformation is prevalent. Social groups are modeled as networks of individuals which can be represented by graphs. Agent-based computer simulations are then used to study how agents' beliefs evolve over time. The authors' findings suggest that when agents discount evidence, – depending on the shape of the network – the speed of convergence, i.e. how fast networks reach true beliefs, is reduced. They argue that understanding these dynamics can ultimately provide insights into how the adverse effects of disinformation can be mitigated in real-world communities.

The section then shifts the focus to strategies for countering the spread of disinformation and misinformation. Among these, the quality rating of media outlets and fact-checking initiatives are usually seen as key approaches. However, their impact is limited, as argued by **Sacha Ferrari** in his paper *War against Fake News: Avoiding Manichaeism.* To substantiate his thesis the author investigates the motivations and reasons why media outlets and fact-checking agencies engage in this activity beyond the stated aim of providing better information and improving rational discourse in democracies. Ferrari shows that there is no consensus in the literature about the assessment of the quality rating of media outlets and the alleged objectivity of fact-checkers. The author concludes that these findings provide a nuanced understanding of the circumstances under which the fight against fake news is undertaken. On this basis, the author eventually challenges the popular polarized picture of fake news producers versus "savior" fact-checkers.

For **Lei Niu** the strategies previously mentioned very likely count as efforts to combat disinformation on an individual level, but not necessarily on a collective level. In his paper *Targets of Countermeasures against Disinformation: Individual Beliefs and Group Beliefs* Niu observes that the impact of disinformation on group beliefs has been largely overlooked in these discussions. He argues that since a group can be analyzed as an epistemic agent, it can hold false beliefs resulting from disinformation even if none of its individual members do. This idea underscores that group belief is not just a cumulative function of the beliefs of its individual members, revealing the limited efficiency of efforts that focus exclusively on the impact of disinformation on individuals.

References

Altay, S., Berriche, M., & Acerbi, A. (2023). Misinformation on misinformation: Conceptual and methodological challenges. *Social Media + Society, 9*(1). https://doi.org/10.1177/20563051221150412

Arnold, M., & Wiengarn, J. (2024). Algorithm-driven reconfigurations of trust regimes: An analysis of the potentiality of fake news. In M. Arnold, B. Prietl, H. Heuer, J. Jarke, Y. Boeva, & S. Egbert (Eds.), *Algorithmic regimes* (pp. 187–206). Amsterdam University Press.

Arntzen, F. (2007). *Psychologie der Zeugenaussage: System der Glaubhaftigkeitsmerkmale* (4th ed.). Beck.

Asch, S. (1951). Effects of group pressure upon the modification and distortion of judgment. In H. Guetzkow (Ed.), *Groups, leadership and men* (pp. 177–190). Carnegie Press.

Asch, S. (1956). Studies of independence and conformity: A minority of one against a unanimous majority. *Psychological Monographs, 70*, 1–70.

Ballantyne, N., Celniker, J. B., & Dunning, D. (2024). Do your own research. *Social Epistemology, 38*(3), 302–317. https://doi.org/10.1080/02691728.2022.2146469

Baptista, J. P., & Gradim, A. (2022). A eorking definition of fake news. *Encyclopedia, 2*(1), 632–645. https://doi.org/10.3390/encyclopedia2010043

Baurmann, M., & Cohnitz, D. (2021). Trust no one? In A. K. Flowerree, T. Grundmann, & S. Bernecker (Eds.), *The epistemology of fake news* (pp. 334–357). Oxford University Press.

Bernecker, S., Flowerree, A. K., & Grundmann, T. (2021). Preface. In S. Bernecker, A. K. Flowerree, & T. Grundmann (Eds.), *The epistemology of fake news* (pp. 1–16). Oxford University Press.

Coleman, J. S. (1990). *Foundations of social theory* (1st ed.). Belknap Press of Harvard Univ. Press.

Coady, D. (2019). The trouble with 'fake news'. *Social Epistemology Review and Reply Collective, 8*(10), 40–52.

Fallis, D., & Mathiesen, K. (2019). Fake news is counterfeit news. *Inquiry*, 1–20. https://doi.org/10.1080/0020174x.2019.1688179

Floridi, L. (2019). The routledge handbook of philosophy of information. In *Routledge handbooks in philosophy*. Routledge.

Franzen, A., & Mader, S. (2023). The power of social influence: A replication and extension of the asch experiment. *PLoS ONE, 18*(11), e0294325.

Gambetta, D. (1988). Can we trust trust? In D. Gambetta (Ed.), *Trust: Making and breaking cooperative relations* (pp. 213–237). Blackwell.

Gelfert, A. (2018). Fake news: A definition. *Informal Logic, 38*(1), 84–117.

Gelfert, A. (2021). Fake news, false beliefs, and the fallible art of knowledge maintenance. In A. K. Flowerree, T. Grundmann, & S. Bernecker (Eds.), *The epistemology of fake news* (pp. 310–333). Oxford University Press.

Greuel, L., Fabian, T., & Stadler, M. (1997). *Psychologie der Zeugenaussage: Ergebnisse der rechtspsychologischen Forschung*. Beltz Psychologie-Verlags-Union.

Grundmann, T. (2023). Fake news: The case for a purely consumer-oriented explication. *Inquiry, 66*(10), 1758–1772. https://doi.org/10.1080/0020174X.2020.1813195

Habgood-Coote, J. (2019). Stop talking about fake news! *Inquiry, 62*(9–10), 1033–1065. https://doi.org/10.1080/0020174x.2018.1508363

Harambam, J., & Aupers, S. (2015). Contesting epistemic authority: Conspiracy theories on the boundaries of science. *Public Understanding of Science, 24*(4), 466–480.

Hardwig, J. (1991). The role of trust in knowledge. *The Journal of Philosophy, 88*(12), 693–708. https://doi.org/10.2307/2027007

Hume, D. (2007). An enquiry concerning human understanding. In P. F. Millican (Ed.), *Oxford world's classics*. Oxford University Press.

Jaster, R., & Lanius, D. (2021). Speaking of fake news: Definitions and dimensions. In S. Bernecker, A. K. Flowerree, & T. Grundmann (Eds.), *The epistemology of fake news* (pp. 19–45). Oxford University Press.

Kaminski, A. (2017). Hat Vertrauen Gründe oder ist Vertrauen ein Grund? Eine dialektische Tugendtheorie von Vertrauen und Vertrauenswürdigkeit. In J. Kertscher & J. Müller (Eds.), *Praxis und 'zweite Natur'. Begründungsfiguren normativer Wirklichkeit in der Diskussion* (pp. 121–139). Mentis.

Kaminski, A. (2021). *Die verwickelte Einfachheit von Vertrauen – und seine spekulative Struktur*. (Unpublished habilitation thesis). Universität Marburg.

Kaminski, A. (2024). Trust in science: A systematic-historical introduction. In M. Resch, N. Formanek, A. Joschy, & A. Kaminski (Eds.), *Trust in science. Science and art of simulation (SAS Edition)* (pp. 1–25). Springer.

Klein, E. (2020). *Why we are polarized*. Profile Books.

Königs, P. (2024). In defense of 'surveillance capitalism'. In *Philosophy & Technology* (Vol. 37, p. 122). https://doi.org/10.1007/s13347-024-00804-1

Moran, R. (2006). Getting told and being believed. In J. Lackey & E. Sosa (Eds.), *The epistemology of testimony* (pp. 272–306). Oxford University Press.

Mößner, N. (2018). Trusting the media? TV news as a source of knowledge. *International Journal of Philosophical Studies, 26*(2), 205–220.

Nguyen, C. T. (2020). Echo chambers and epistemic bubbles. *Episteme, 17*(2), 141–161. https://doi.org/10.1017/epi.2018.32

Nguyen, C. T. (2021). How twitter gamifies communication. In J. Lackey (Ed.), *Applied epistemology* (pp. 410–436). Oxford University Press.

Pennycook, G., & Rand, D. G. (2021). The psychology of fake news. *Trends in Cognitive Sciences, 25*(5), 388–402.

Pennycook, G., Epstein, Z., Mosleh, M., Arechar, A. A., Eckles, D., & Rand, D. G. (2021). Shifting attention to accuracy can reduce misinformation online. *Nature, 592*(7855), 590–595.

Stern, W. (Ed.). (1903). *Beiträge zur Psychologie der Aussage: Mit besonderer Berücksichtigung von Problemen Der Rechtspflege, Pädagogik, Psychiatrie und Geschichtsforschung*. Johann Ambrosius Barth Verlag.

Tandoc, E. C., Jr., Wei Lim, Z., & Ling, R. (2018). Defining 'fake news': A typology of scholarly definitions. *Digital Journalism, 6*(2), 137–153. https://doi.org/10.1080/21670811.2017.1360143

Wiengarn, J. (2022). *Die Grammatik des Vertrauens: Eine Untersuchung in interpersoneller und epistemischer Hinsicht*. Kölner UniversitätsPublikationsServer (KUPS). https://kups.ub.uni-koeln.de/64958/

Wiengarn, J. (n.d.). Forthcoming. Wer Ist hier eigentlich Fake News? Automatisierte Fake News Erkennung und die Politik des öffentlichen Raums. In *Jahrbuch Technikphilosophie*.

Wittgenstein, L. (2015). In G. E. M. Anscombe & G. H. von Wright (Eds.), *Über Gewißheit* (13th ed.). Suhrkamp.

Varieties of Disinformation

Trust in Falsehoods: Definitions, Sources, Challenges

Andrija Šoć

Contents

Abstract This paper consists of three parts. In the first part, I offer a conceptual analysis of the terms most commonly associated with falsehoods – misinformation, disinformation, and malinformation. I explore different ways in which this trichotomy has to be defined more precisely for us to understand the effects of each type of falsehood. In the second part, I provide a comprehensive definition of trust, which I then proceed to apply to different types of falsehoods. I outline four different sources of trust and try to match each to a complex taxonomy within what I call the 'landscape of deception'. In the third part, I try to offer a broad suggestion for how standard approaches to solving the problem of trust in falsehoods could be improved in light of the preceding analysis. Finally, I try to show that an even greater concern potentially pertains to the appearance of different generative AI models.

A. Šoć (✉)
Faculty of Philosophy, Institute of Philosophy, University of Belgrade, Belgrade, Serbia
e-mail: andrija.soc@f.bg.ac.rs

M. M. Resch et al. (eds.), *Trust and Disinformation*,
https://doi.org/10.1007/978-3-031-96790-0_2

1 Falsehoods: Concepts and Definitions

The increased ideologization and polarization of societal issues brought about a gamut of different types of false information: disinformation, misinformation, malinformation, fake news, rumors, and propaganda.[1] However, these phenomena are mutually distinct, and to conflate them has two problematic consequences. First, we lose the ability to distinguish between different ways in which news recipients come to form false beliefs on crucial issues. Second, we fail to understand different sources of trust in different forms of such falsehoods.

The goal of this paper is to attempt a rectification of both these issues. I will attempt to do this in three stages. First, I will reexamine the standard definitions of different types of falsehoods. The most common types are misinformation, disinformation, and malinformation. I will revisit two slightly different attempts to define all three and show that delineating them is more complicated than recognized within extant definitions. Providing conceptual clarification on these three types of falsehoods will help us set up a broader set of falsehoods, which will also include, among other items, fake news, spin, and propaganda.

Second, I will apply a comprehensive definition of trust, which I formulated in an earlier paper (Šoć, 2021a, 2021b). Since the definition is a bit more complex than those usually mentioned in the literature,[2] one way to justify it will be to use it to explain forms of trust in received falsehoods. In order to accomplish that, at the second stage of the argument, I will discuss the types of falsehoods mentioned in the first paragraph.

The next stage of my attempt is to create a map of these different categories and combine it with the key aspects of trust mentioned in the definition: cognitive, affective, evaluative, and motivational. As I will try to show, there are different sources of trust in each of these categories, and identifying the particular source of trust will, as I try to show in the third part of the paper, help us offer contextually appropriate measures of protection against believing in statements belonging to each of these types of falsehoods.

I begin with the most frequent suspects within the realm of falsehoods – misinformation, disinformation, and malinformation. According to UNESCO's "Journalism, Fake news and Disinformation" (Ireton & Posetti, 2018, 43–54),[3] 'misinformation' implies that those who disseminate it believe in its contents, 'disinformation' pertains to knowing the spread of false information, whereas 'malinformation' implies that those who disseminate it know that it is false and have a conscious intent to deceive the recipients of the message by combining the

---

[1] In this paper, I will refer to these as types or instances of falsehoods. Considered jointly, I treat them as different categories within what I call 'the landscape of deception'.

[2] See, e.g. Hardin, 1992 and Newton et al., 2018. In Šoć (2021a, 2021b), I critically deal with these and some additional proposals.

[3] See: https://en.unesco.org/sites/default/files/journalism_fake_news_disinformation_print_friendly_0.pdf. Accessed 13 Mar 2023.

falsehood with elements of truth. Unlike disinformation, then, malinformation doesn't merely contain outright falsehoods. Rather, it is couched in something true, but is taken out of context or is incorrectly extrapolated, for instance.

Another way to approach the three is to follow (Wardle & Derakhshan, 2017, 20).[4] They consider misinformation, disinformation, and malinformation to be examples of 'information pollution' or 'information disorder' (ibid). Their definitions go as follows:

> [1] Dis-information: Information that is false and deliberately created to harm a person, social group, organization or country, [2] Mis-information: Information that is false, but not created with the intention of causing harm, [3] Mal-information: Information that is based on reality, used to inflict harm on a person, organization or country. (Wardle & Derakhshan, 2017, 5).

The visual representation which accompanies this definition shows that the two key criteria of sorting the three are whether they are false and whether they are harmful. In addition, they also mention intention as a relevant definitional element.

Even though at face value such detailed criteria are far more helpful in probing how different types of falsehoods work, and how to distinguish between these three, there still are problems we need to further explore. Namely, the distinguishing feature of misinformation is that the falsehoods shared by individuals or organizations are disseminated with the belief that they are true, and without the intention to cause harm. On the other hand, misinformation and disinformation are more akin to being entirely false, whereas malinformation contains at least some truths (distorted or otherwise decontextualized).

We can see where the problem with the basis of such distinction lies if we first ask whether the definitions provided are wide or narrow. Second, we will see that the elements that do make up the proper *fundamentum divisionis* require more nuance than has been provided. The answer to the first question is that these definitions are too narrow. In other words, they fail to account for certain types of informational outputs (we can use the generic term 'utterance'), which contain different ordering of definitional elements. For instance, one can find utterances that are, unlike in the case of misinformation (as defined), knowingly false, but are not meant to do harm, and might even be provided to benefit the utterer. Think of a sales call or any number of different types of advertisements. When products are over-marketed, or benefits over-stated, it is not always intended that harm be caused to the consumer. Although the harm may come from a poorly designed product, in many cases the product's effects are simply neutral in terms of harm inflicted, or in terms of any particular benefits.

Another way the above definitions are inadequate is that they do not capture utterances that are knowingly disseminated as false, but only for the benefit of the utterance target – be it an individual, a group, or the entire public. Think of any typical example of a 'white lie', or an instance in which falsehood is thought to produce

[4] Full report at: https://rm.coe.int/information-disorder-toward-an-interdisciplinary-framework-for-researc/168076277c. Accessed 14 Mar 2023.

a beneficial consequence. One prominent example of such behavior was the situation at the beginning of the pandemic when different public statements about wearing masks and gloves rather reflected the belief that the public will react in a way that will bring not harm to anyone but rather have a more positive outcome than an utterance which was at the time considered to be true.[5] Specifically, the issue of masking was at first viewed in terms of product availability, which is why, in the words of the utterers themselves, the message to the public was that masking doesn't work so that masks wouldn't become deficient at the time when the hospital staff genuinely needed them.[6] Regardless of any further implications of such an act, and the effect on public trust it had,[7] it is clear that the above definitions fail to adequately capture such a category.

One further worry is that they seem to imply that the one way to combat information disorder is to always utter truths. However, regardless of the admirable sentiment behind that statement, the political and societal realities are far more complex. What seems to be left out of such a sentiment is the fact that truth can also be harmful to individuals, groups, the public, or a state. This is why there are laws against revealing documents with 'top secret' designation, espionage and industrial espionage, and the like. Such instances also can be argued to contribute to the phenomenon the authors above termed 'information' pollution or 'information disorder'. The complexity of such a landscape of issues can also contain multiple stages. Thus, revealing some truth at a certain point in time can later be used to spin this, recontextualize it, or create fertile grounds for spreading harmful malinformation. To successfully combat the informational harm, we would need a more nuanced way to sort the three key types of falsehoods we examined in the past few paragraphs. I propose we do that in the following steps.

First, instead of mentioning three differentiating criteria – intention, falseness, and harm – I use four – intention, knowledge, falseness, and harm. Namely, introducing the criterion that pertains to one's knowledge (or lack thereof) has two primary benefits: it provides nuance to our analysis of falsehoods as usually defined while also pointing to the need for better definitions; it also helps us understand how falsehoods can target different aspects of our trust attitudes. To that end, I will provide a comprehensive definition of the concept of trust at the beginning of next section. Second, I group the criteria by whether they pertain to the subject or the object. Intention and knowledge belong to the former, and the falseness and harm to

[5] An example of a statement of this sort can be found at: https://edition.cnn.com/factsfirst/politics/factcheck_e58c20c6-8735-4022-a1f5-1580bc732c45. 17 Mar 2023.

[6] To quote from the fact-checking explanation linked in the previous footnote: "While Fauci, along with several other US health leaders, initially advised people not to wear masks, Fauci later said that he was concerned that there wouldn't be enough protective equipment for health care workers. This was also early in the pandemic before public health experts fully knew how contagious the disease was and how it spread."

[7] See, e.g. https://www.frontiersin.org/articles/10.3389/fpsyg.2021.648692/full. Accessed 18 Mar 2023. The early shift in public trust on CDC and its officials has been touched upon in (Šoć, Jovanović, 2024).

the latter. However, as the four are still mutually distinct, further specification is needed. Thus, three, there is something objective regarding the truth value of an utterance. It is either true or false. Four, we can wonder whether a subject actually knows or can know if the utterance is true or false. This element is distinct from the intention and can, as we have seen from the examples above, vary in ways not covered by original definitions. Finally, the effect of sharing an utterance – broadly, harm or benefit (though these aren't the only two options) – needs to be assessed in different ways, depending on the context, as well as diachronically. Namely, something that brings immediate benefit can cause long-term harm and vice-versa. For instance, a cluster of analyses would condemn early statements about masks, because the initial benefit (no mask shortages) was replaced by long-term harm (loss of trust in public officials).[8]

One final issue to consider with the usual definitions of misinformation, disinformation, and malinformation is that too often they neglect one key question: can 'information' itself be true or false, or should the term refer only to something true? In other words, could there be false information? Even more succinctly: is it information if it is false? Because there exist two different domains within which we set our four criteria – the domain of agents or subjects and the domain of utterances, information, or objects – Fig. 1 is an inadequate representation of the relation between misinformation, disinformation, and malinformation. Instead, one continuum pertains to truth or falsity, and it includes something true, something that is false, and a truth combined with falsity. The other continuum pertains to knowledge

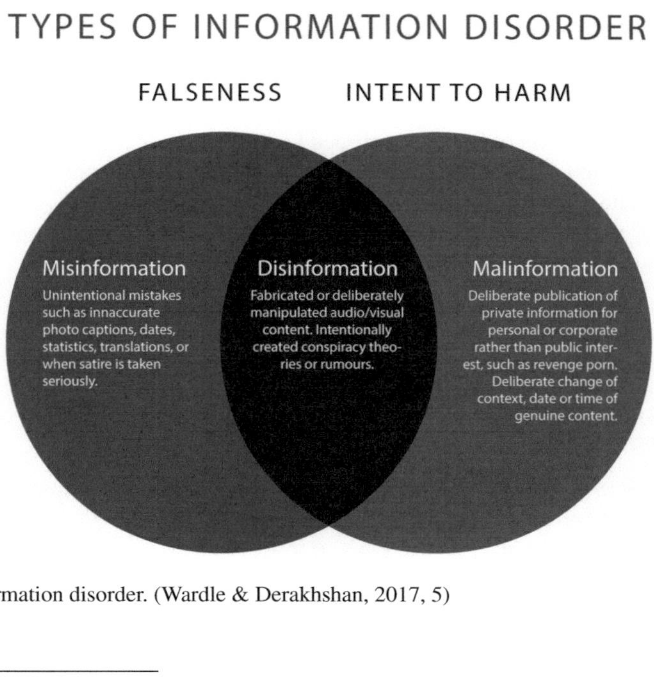

Fig. 1 Information disorder. (Wardle & Derakhshan, 2017, 5)

[8] See, e.g., the article at footnote 7.

and intention. Since it is difficult to always quantify harm, and it can come from either true or false utterances, I propose we do not include it when sorting different types of falsehoods but keep it prominent when explaining each type of falsehood and its effect. In other words, the use of the criterion of harm is not definitional, but explanatory.

Furthermore, since the term 'false information' can be thought of as a contradiction in terms, this continuum will include information, disinformation, and malinformation. On the other hand, because misinformation is primarily defined by the epistemic state of the agent (lack of knowledge and lack of deceptive intention), it will be on a different continuum. This will include misinformation, but not disinformation or malinformation. In fact, the entire landscape will have to be thought of slightly differently than usual, to capture all the nuances we described. It will be important to capture different ways in which different types of falsehoods affect their targeted audience. Because the definitions of the misinformation, disinformation, and malinformation are problematic in ways I suggested above, I will try to show that a broader scope of falsehood categorization will better correspond to the broad scope of ways in which one may find falsehoods trustworthy. Thus, I propose to include the following in our landscape: misinformation, disinformation, malinformation, fake news,[9] conspiracy theories, rumors, alternative facts, spin, and propaganda. This is by no means an exhaustive list, but rather an attempt to cover the ground in more detail. I now turn to the exploring the sources of trust in these types of falsehoods.

2 Sources of Trust

I begin with a definition of trust I developed in an earlier paper (Šoć, 2021a, 2021b). I will not reiterate my criticism of different alternative conceptions of trust, such as trust being understood as behavior, as expectation, or as a vaguely defined psychological state, failing to adequately capture the complexity of the phenomenon. To quote the definition[10] again:

> Trust is an attitude that a citizen has with respect to a certain person, politics, institution or society, partly constituted by the conditions in which the citizen was born and is living now, determined in part by rational expectations from the object of trust, in part by the affective profile, and in part by the values the citizen holds; this attitude directs the citizen to act in a certain way, and renders the citizen inclined to make a certain choice when the opportunity arises. (Šoć, 2021a, 2021b, 49).

The main upshot of accepting this definition is that it tries to encompass four different elements of our complex attitude of trust: cognitive, affective, motivational/behavioral, and evaluative. The best way to test such a definition is to try and apply

[9] See more on this topic in, e.g.: Giusti & Piras, 2021 (eds.).

[10] It was originally provided in Serbian.

it to concrete phenomena, which is why if we can allocate different trust attitudes to different types of falsehoods, we will be able to both affirm the use of the given definition, but also provide a more nuanced explanation of how falsehoods really capture the attention of individuals or the public as a whole.

In the broadest of strokes, we can initially note that some types of falsehoods appeal more to our rational side, and some to our emotions (fears, hopes, etc.). Some target our value structures, whereas some directly portend to render us active or inactive. The source of our trust in them will also play a salient role in our reception and further dissemination, as well as in our tendency to re-check our sources and question whether something that appeals to, say, our affective or evaluative attitudinal structures is indeed true or is perhaps false. It is important to also note here that because the attitude of trust is so complex, it may rarely happen that just one type of trust attitude will be involved. However, we can still determine the dominant ones, which I will return to in the third part of the paper when discussing strategies for combating trust in falsehoods.

Listing again the landscape of falsehoods: misinformation, disinformation, malinformation, fake news, conspiracy theories, rumors, alternative facts, spin, and propaganda, we can start from the initial trichotomy of misinformation, disinformation, and malinformation. Having in mind our discussion in the first part of the paper, we can better discern why it is that agents trust in utterances denoted by these terms. Extended definitions of each of the three points toward how to understand our trust in them. Malinformation combines truth and falsity, which indicates that it targets the cognitive element of trust. In other words, it appeals to our prior knowledge, or our inclination to consider ourselves informed, in order to generate our trust in it. On the other hand, because misinformation is disseminated without the intention to cause harm, or knowledge that it is false, we can infer that a subject's willingness to share such content or to accept it resides more often in the affective element of trust. For instance, a person afraid of the effects of the pandemic and in search of a cure will further disseminate a piece of news that indicates that a medicine like ivermectin or chloroquine prevents or treats Covid-19. This type of case can help us see how the phenomenon of trust in disinformation, or in falsehoods as such, can be properly analyzed using our definition. We can say that the deception behind disinformation appeals, just like misinformation, to our affective state. But, we can go a step further.

Namely, we can now introduce other types of falsehoods. Because harm and falseness are the defining features of 'disinformation', we can broaden our analysis to include fake news and trust in it, or spin and rumors,[11] which are more prominent in a public setting when political struggle (and especially polarized election campaigns) is involved. We can say that rumors appeal more to the cognitive and the affective component of trust than to the evaluative or behavioral, and that spin is an attempt to deceive the public by taking into account the cognitive attitude of news

[11] An excellent treatment of rumors from the perspective of applied epistemology can be found in, e.g.: Coady, 2012 and Gelfert, 2019.

consumers. Fake news, on the other hand, frequently features polarizing content that aims to capture the affective and evaluative profile of news consumers. Thus, subjects who are inclined to believe in a fake piece of news and are not eager to check other sources or impartial agencies, mostly exhibit a confirmation bias.[12] Now, if a conservative or pro-life news organization said that research shows that embryos can feel pain one week after conception, they might be inclined to trust this utterance and fail to check whether other news sources report this or whether such research exists. In such a case, the primary source of trust is the system of values which is unshakeable and thus makes the utterance, for this particular subject, immune to re-examination. The source is hardly cognitive or behavioral, but will to some degree also be affective.

We have so far mentioned misinformation, disinformation, malinformation, fake news, spin, and rumors. We haven't gone into a detailed analysis of this taxonomy, as it would take far too much space. However, what we have seen is that dominant sources of trust vary with respect to these different types of false utterances. Malinformation and spin are characterized by cognitive elements of trust, misinformation, and rumors by affective, disinformation by a combination of cognitive and affective, and fake news by a combination of affective and evaluative components of the trust-attitude. Two other types of falsehoods we still haven't discussed in this context are conspiracy theories[13] and propaganda.[14] These two are, it seems, far more complex in structure than the other mentioned types, and are also broader in scope and effect. Let us examine them more closely, with the caveat that our context is only the source of trust, not these phenomena in themselves, as this would require a far more detailed investigation.

In terms of trust in conspiracy theories, various examples can tell us that the main components of trust are cognitive and behavioral, with secondary elements of affective and evaluative components also being present. Most of the conspiracy theories, be it the denial of the Moon landing, the existence of UFOs, or belief in various secret societies which control entire governments,[15] rely on an extensive list of supposed facts and personal reports of authoritative figures.

Perhaps the most insidious and far-reaching effects within the landscape of deception are produced by propaganda. Because of its context, where the perpetrators of falsehoods are not individuals but ambitious organizations or entire powerful states, propaganda's multifaceted nature cannot be captured within one section or one paper. However, within the confines of our discussion, we can point squarely to the sources of trust in propaganda. Namely, like conspiracy theories, it elicits a complex set of trust-related attitudes. Also, like conspiracies, the trust in it elicits motivation and behavior that reveals the acceptance of the transmitted message.

---

[12] See, e.g.: Kholekhile et al., 2022.

[13] See extensive discussions of conspiracy theories in: Coady, 2006 (ed.) and Coady, 2012.

[14] A detailed source for research on propaganda can be found, e.g. in: Auerbach and Castronovo, 2013 (eds.) and Stanley, 2015.

[15] See, e.g.: Harambam, 2020.

However, unlike a conspiracy theory, a piece of propaganda aims to dominantly elicit an affective response and it targets a person's system of values. Thus, the dominant components of trust are affective, evaluative, and behavioral/motivational, with the cognitive component being only incidentally present.

Our discussion thus far has, as I tried to show, revealed a richly textured spectrum of falsehoods and a complex phenomenon of trust, featuring in a dynamic multifaceted interplay. Tentatively, we may sum up our discussion in the following table:

Note that different hues are used to indicate the dominance of different sources of trust. A darker hue indicates dominance or salience, and a lighter hue the relevance, though not typical or dominant presence, of a particular component of trust-attitude. Naturally, for an exact analysis, one can only examine each piece of falsehood in detail to determine how it elicits trust. This, too, leads us to the further point and the ultimate goal of this paper – exploring the relevant factors for formulating strategies that would help individuals, organizations, and the general public to combat attempts at disseminating falsehoods.

Namely, rather than featuring one set of prescriptions aimed at vaguely becoming more informed, trusting the facts presented by reliable organizations, etc., stakeholders need to take into account the variance that exists within the landscape of deception, and the different trust attitudes that track different types of falsehoods within this landscape. One additional benefit of such an approach is that it can be used to discern the rising issue of innovative technologies, such as the AI models, which are about to reshape an entire informational space, and need to be fully taken into account.

3 Falsehoods, AI, and the Informational Paradox

The metaphor for the entire gamut of different types of falsehoods was termed in this paper 'the landscape of deception'. As this landscape becomes more and more populated, and the proliferation of falsehoods speeds up, so different strategies for combating such phenomena also proliferate. I will not try to examine particular strategies or suggest some radically new ones. Instead, I will try to apply the upshots from the previous two parts to answer one specific question: "How should a strategy be formulated in order to successfully combat falsehoods?". I will turn to a claim put forward in the recent book by Bakir and McStay (2022) dealing with this topic so that I can both situate my approach and try to show how an analysis based on the particularized notion of sources of trust in different types of falsehoods represents a necessary refinement for ultimate formulations of successful informational strategies.

In their book, Bakir and McStay correctly summarize different standard strategies for combating misinformation and disinformation. They mention "providing guidelines for journalists", "greater journalistic transparency", "integrating source transparency indicators into ranking algorithms", and "developing technical specifications", among other suggestions (Bakir and McStay, 2022, 224–225). However, they point out that "these solutions also do nothing to address the root cause of distrust in news, namely, the long-standing political and commercial processes of manipulation and commodification". (ibid, 225). Their further suggestions indicate seven major areas of providing solutions to the problem of false information – governmental action, cybersecurity, digital platforms and intermediaries, advertising, professional persuaders and public relations in the political domain, media organizations, and education[16] – are necessary, but thus far ineffective (ibid., 234). According to Bakir and McStay, "the ultimate solution would be to alter the business models of platforms, so that they do not seek maximal user engagement and so that they do not design algorithm that make emotional and deceptive content go viral" (2022, 234).

While admirable in both ambition, scope, and cohesiveness, both the summation of the extant strategies and the proposal for a more successful alternative ultimately seem to betray an underlying assumption – that to find a successful way to limit the spread of falsehoods one must somehow intervene ultimately on the side of the content, rather than the consumer of the content (even though education campaigns are, as mentioned, a necessary though insufficient, or at least ineffective part of the solution). The place where my assessment diverges from theirs is in the way I see the "root cause" of distrust in news. The complex interplay of aspects of trust and different types of falsehoods which can in different ways appeal to us requires differentiated analysis that takes into account all the different sources of trust correlated with each different type of falsehood. Even then, such an analysis is merely a starting point, with further steps requiring additional subdivisions of each type of falsehood, and research into how each source of trust is targeted is specified across the entire landscape.

The assumption behind this differentiated approach is that trust in falsehoods is not incidental, temporary, or new. It is not essentially grounded in disinformation campaigns or media manipulation, although it is exacerbated by them. Fundamentally, the problem of combating various types of falsehoods has to begin at the conceptual level, at answering basic philosophical questions about reasons for trust and distrust, and at treating the landscape of deception as a spectrum of different opportunities not only for manipulation but also for establishing the epistemic limits of engagement with information sources. This is the starting point and the definition of the concept of trust can provide the first building block for this analysis, as it encompasses a variety of elements of our trust attitudes.

If this is so, then what we are left with are individual citizens, searching for ways to find the reliable sources of information, either confused, skeptical, or firmly entrenched within the viewpoint constantly reinforced by a particular source of their choice. The strategy I am here suggesting, but the one which needs to be developed in a work of a far wider scope, is to combat different types of falsehoods (as listed in Table 1) by using bottom-up approaches. There is ample literature on such

[16] See esp.:Bakir and McStay 2022, 232–234). See also: Gregor and Mlejnková, 2021 (eds.).

Table 1 The sources of trust in different types of falsehoods

Falsehoods/Trust	Cognitive	Affective	Evaluative	Motivational
Misinformation				
Disinformation				
Malinformation				
Fake news				
Spin				
Rumors				
Conspiracy theories				
Propaganda				

approaches in the domain of politics, pertaining to deliberative and participative models of democracy.[17] Whether such bottom-up strategies will be successful remains to be seen. One, at least *prima facie*, advantage it has is that it takes into account both the sources of falsehoods and the targets of the falsehood. It is, thus, more comprehensive and, because of its bottom-up approach, also more specific.

However, one particular issue that seems to slowly appear on the horizon has the potential to either make these strategies either more difficult (or easier) to apply – I refer to the issue of using AI in our engagement with online content.[18] More specifically, the version of AI I refer to is better described as a large-language AI model.[19] While the version of the AI model referred to in literature and media during the writing of this paper is ChatGPT-3, the latest development saw the unveiling of an

[17] This is too broad a topic to be discussed here in detail. I cover some of that terrain in Šoć (2021a, 2021b) and Šoć, 2024. For a comprehensive overview of different approaches within deliberative theories, see, e.g.: Bächtiger et al., 2018.

[18] See, e.g.: Bakir and McStay 2022, 259–261). While the version of the AI model they refer to is GPT-3, as of writing this paper, the latest development saw the unveiling of an improved version, GPT-4. Early indications suggest that it is vastly improved in contrast to the GPT-3. Whether it is improved in the domain of combating disinformation (or will rather facilitate it), remains to be seen.

[19] Perhaps a seminal work that introduces the GPT-3 model is "Language Models are Few-Shot Learners" (Brown et al., 2020).

improved version, GPT-5.[20] Trained on terabytes of text (Wiggers, 2022), this model has shown a remarkable range of uses in just a few months after its unveiling.[21]

Now, the issue I am referring to is not so much connected with the different uses or drawbacks of such a tool. What I have in mind is a more abstract question. Namely, how are we to consider the information received from the AI? The first problem is that it isn't sentient, and especially that it does not possess any particular beliefs or agenda, the way those who originate or spread different types of falsehoods possess. The second problem is entirely metaphysical: there is no 'it' there at all. In fact, a GPT model of AI is not a subject in any sense of the word, but rather a tool that generates text based on a user's prompt.[22] No other models currently or formerly in development can be ascribed some substantive character either. As the 'GPT' term suggests, the text is generated and the generated content is based on the texts the AI model (or a chatbot) is trained with. Thus, if a falsehood is generated by such a model, who is its author? Since the generation is based on a statistically-determined pattern, the author cannot be the chatbot. One easy answer might be that the ones who have programmed the AI are the authors, but even the developers of such a model cannot really know which outputs will follow from which inputs. They provide some constraints on the model, and which prompts it can or cannot respond to, but the control is not complete, nor entirely transparent. As (Rudin & Radin, 2019, 1) describe this issue:

> Over the last few years, the advances in deep learning for computer vision have led to a widespread belief that the most accurate models for any given data science problem must be inherently uninterpretable and complicated. [...] In machine learning, these black box models are created directly from data by an algorithm, meaning that humans, even those who design them, cannot understand how variables are being combined to make predictions.

Based on these passages and previous considerations, there appears to be a certain informational paradox: messages are given with no messenger existing. When accurate or false statements become the result of a user's prompt, they become utterances with no utterer behind them. The problems of trust in falsehoods that are generated in this way seem to be even more pernicious than the problems of trust in falsehoods the public usually encounters, for at least two reasons. First, when formulating a strategy to combat the spread of falsehoods, one naturally strives to propose solutions that affect both the source and the end-users. But if the trust

[20] The name of the model stands for: "Generative Pre-Trained Transformer". For a more accessible introduction, see: Hughes, 2023. Even though technology has improved in the past few years, the fundamental questions raised in the paper remain, precisely due to the generative character of this model, as well as any other large-language model.

[21] One useful, though by no means exhaustive review can be found at: Sundar, 2023.

[22] See one accessible explanation of this claim at Eisikovits, 2023. See also: Hughes, 2023.

placed in AI-generated content cannot refer to the 'author' of a message, then any proposed solution will be at best ineffective, and at worst might directly contribute to the problem. Second, the genuine sophistication of the mechanism of content generation, reflected primarily in humanlike responses, might to a sizeable portion of users appear to yield increased reliability behind the prompted content.[23] This could increase the strength of the trust placed in such a mechanism and attempts to demonstrate that the generated content is false might turn out to be even more difficult than in 'ordinary' cases. Furthermore, the unpredictable combination of impressive and accurate output, with random and inaccurate performance could in a sizeable number of cases throw into doubt the positive effects of deploying generative AI models.

Precisely because of the growing number of difficulties in reaching reliable news sources, attaining trustworthy information, and making that trust well-founded, the need for applying a multifaceted dynamic framework for analyzing types of falsehoods and sources of trust becomes especially important. As discussed at page 13 above, and throughout the paper, the dynamic nature of the framework lies in the recognition that there are constant changes within different groups targeted by falsehoods, and that even within individual members of any particular group the trust attitudes can develop in different directions. The multifaceted nature of the framework, grounded by the definition of trust provided in section "Sources of Trust", helps us correlate what it is about different types of falsehoods that appeals to different groups and individuals (as illustrated in Table 1).

Further development and application of this framework will require analytical segmentation of the entire landscape of deception that was broadly categorized in this paper. For this analysis to yield successful strategies, both within the context of traditional mainstream media, established digital and social media, and AI-driven content generation, it is paramount to give up the attempt to find one overarching approach. Complex solutions that would play an effective role in resolving complex informational challenges will require extensive collaboration within and across a wide range of scientific fields. The discussion in this paper represents a broad attempt to outline the scope of challenges ahead of us. Hopefully it also provides a set of signs that point to a promising way ahead.

Acknowledgment The realization of this research was financially supported by the Ministry of Science, Technological Development, and Innovations of the Republic of Serbia, as a part of financing scientific research at the University of Belgrade, Faculty of Philosophy (contract number 451-03-66/2025-03/200163). I thank the reviewer of this paper for insightful comments.

[23] See, e.g., a recent study by Spitale et al. (2023).

References

Auerbach, J., & Castronovo, R. (2013). *The oxford handbook of propaganda studies*. Oxford University Press. https://doi.org/10.1093/oxfordhb/9780199764419.001.0001

Bächtiger, A., Dryzek, J. S., Mansbridge, J., & Warren, M. (2018). *The oxford handbook of deliberative democracy*. Oxford University Press. https://doi.org/10.1093/oxfordhb/9780198747369.001.0001

Bakir, V., & McStay, A. (2022). *Optimising emotions, incubating falsehoods: How to protect the global civic body from disinformation and misinformation*. Palgrave Macmillan.

Brown, T., et al. (2020). Language models are few-shot learners. *Advances in Neural Information Processing Systems, 33*, 1–25.

Coady, D. (Ed.). (2006). *Conspiracy theories: The philosophical debate*. Ashgate.

Coady, D. (2012). *What to believe now*. Blackwell.

Eisikovits, N. (2023). AI isn't close to becoming sentient – the real danger lies in how easily we're prone to anthropomorphize it. In *The conversation*. Retrieved March 18, 2020, from https://www.yahoo.com/news/ai-isnt-close-becoming-sentient-122254552.html

Gelfert, A. (2019). Rumors. In D. Coady & J. Chase (Eds.), *The routledge handbook of applied epistemology* (pp. 259–271). Routledge.

Giusti, S., & Piras, E. (Eds.). (2021). *Democracy and fake news: Information manipulation and post-truth*. Routledge.

Gregor, M., & Mlejnková, P. (Eds.). (2021). *Challenging online propaganda and disinformation in the 21st century*. Palgrave Macmillan.

Harambam, J. (2020). *Contemporary conspiracy culture: Truth and knowledge in an era of epistemic instability*. Routledge.

Hardin, R. (1992). The street-level epistemology of trust. *Analyse & Kritik, 14*, 152–176. https://doi.org/10.1515/auk-1992-0204

Hughes, A. (2023). ChatGPT: Everything you need to know about OpenAI's GPT-4 tool. In *ScienceFocus, March 16*. Retrieved March 19, 2020, from https://www.sciencefocus.com/future-technology/gpt-3

Ireton, C., & Posetti, J. (Eds.). (2018). *Journalism, 'fake news' & disinformation: Handbook for journalism education and training*. United Nations Educational, Scientific and Cultural Organization.

Kholekhile, L. G., Wang, J., & Zifla, E. (2022). Can warnings curb the spread of fake news? The interplay between warning, trust and confirmation bias. *Behaviour & Information Technology, 41*(16), 3552–3573. https://doi.org/10.1080/0144929X.2021.2002932

Newton, K., Stolle, D., & Zmerli, S. (2018). Social and political trust. In E. M. Uslaner (Ed.), *Oxford handbook of social and political trust* (pp. 37–56). Oxford University Press.

Rudin, C., & Radin, J. (2019). Why are we using black box models in AI when we don't need to? A lesson from an explainable AI competition. *Harvard Data Science Review, 1*(2), 1–10. https://doi.org/10.1162/99608f92.5a8a3a3d

Spitale, G., et al. (2023). AI model GPT-3 (dis)informs us better than humans. *Science Advances, 9*(26), eadh1850. https://doi.org/10.1126/sciadv.adh1850

Stanley, J. (2015). *How propaganda works*. Princeton University Press.

Šoć, A. (2021a). Ka obuhvatnom pojmu poverenja [Toward a comprehensive concept of trust, in Serbian]. *Kritika, 2*, 35–53. https://doi.org/10.5281/zenodo.4784241

Šoć, A. (2021b). Deliberative education and quality of deliberation: Toward a critical dialogue and resolving deep disagreements. In I. Cvejić, P. Krstić, N. Lacković, & O. Nikolić (Eds.), *Liberating education: what from, what for?* (pp. 123–146). University of Belgrade, Institute for Philosophy and Social Theory.

Šoć, A. (2024). Participation, responsiveness, interaction: The benefits of epistemic checks-and-balances for the revitalization of democratic institutions. In I. Fiket, Č. Markov, V. Ilić, & G. P. Draško (Eds.), *Participatory democratic innovations in southeast europe: How to engage in flawed democracies* (pp. 37–58). Routledge.

Šoć, Andrija and Monika Jovanović. 2024. Only a theory? Substantive and methodological strategies for regaining trust in science. (forthcoming in: The science and art of simulation trust in science, eds. Michael M. Resch, Nico Formanek, Ammu Joshy, Andreas Kaminski. Springer, 2024).

Sundar, S. (2023). If you still aren't sure what ChatGPT is, this is your guide to the viral chatbot that everyone is talking about. In *Business Insider*. Retrieved March 22, 2023, from https://www.businessinsider.com/everything-you-need-to-know-about-chat-gpt-2023-1

Wardle, C., & Derakhshan, H. (2017). Information disorder: toward an interdisciplinary framework for research and policy making. In *Council of Europe*. Retrieved March 14, 2023, from https://rm.coe.int/information-disorder-toward-an-interdisciplinary-framework-for-researc/168076277c

Wiggers, K. (2022). *The emerging types of language models and why they matter*. TechCrunch. Retrieved March 24, 2023, from https://techcrunch.com/2022/04/28/the-emerging-types-of-language-models-and-why-they-matter

When Homeopathy Cures Cancer: About Bullshit, Irony and Disinformation in the Homeopathic Discourse

Viet Anh Nguyen Duc and Marcel Erik Lemmer

Contents

Abstract This paper distinguishes between different forms of disinformation in the context of the homeopathy discourse. To this end, the paper first looks at the motives of the various narrative carriers (True Believers, Committed persons, Bullshitters, Semi-Ironic persons, and Entrepreneurs). We argue that the type of disinformation depends on the motives. We differentiate between a strategic and an expressive modality of disinformation. In the strategic modality, false information is spread for the sake of deception, for example, to convince a person of a specific cause or to encourage them to buy a product. In the expressive modality, false news is spread to articulate mistrust of established institutions (science, governments, conventional medicine, etc.). This paper aims to concretize this distinction in more detail by examining the aforementioned narrative carriers and to work out further internal differentiations.

V. A. N. Duc
TU Darmstadt, Darmstadt, Germany
e-mail: viet_anh.nguyen_duc@tu-darmstadt.de

M. E. Lemmer (✉)
Karlshochschule International University, Karlsruhe, Germany
e-mail: mlemmer@karlshochschule.org

M. M. Resch et al. (eds.), *Trust and Disinformation*,
https://doi.org/10.1007/978-3-031-96790-0_3

1 Introduction

The discourse surrounding homeopathy as the most significant challenger to hegemonic medical practice has become exacerbated and polarized. Almost the entire academic medical community declares homeopathy to be completely ineffective. They argue homeopathic procedures have no demonstrable therapeutic effect beyond placebo and contextual effects (Lübbers & Endruscheit, 2021, 682). Homeopathy has been subject to this criticism since its inception (Jütte, 1996, 19). Supporters of homeopathy, on the other hand, speak of a highly effective alternative (complementary) medicine (Gebhardt, 1991, 81). "The one who cures is right" is a self-confident slogan offered by homeopaths.[1] With this phrase, the homeopathic community claims to have cured a large number of people with various ailments. Anyone who claims to have cured anything from autism[2] to cancer[3] is certainly not hiding.

On the other hand, critics such as the *Society for the Scientific Investigation of Parascience* ("Gesellschaft zur wissenschaftlichen Untersuchung von Parawissenschaften") refuse to acknowledge any of these healing achievements. They warn against substituting effective conventional treatment with ineffective homeopathic remedies and highlight their concerns with high-profile campaigns.[4] Their critical standpoint points out that homeopathic remedies are irrational procedures in which no active ingredients can be detected due to their extreme dilution. Scientific mimicry (Mukerji & Ernst, 2022, 10) and deliberate deception (Lübbers & Endruscheit, 2021, 685) are at issue. The controversy surrounding homeopathy, therefore, raises questions not only about its medical efficacy but also about the quality and integrity of the information. Disinformation is a particularly problematic phenomenon in this context, as it is based on a deliberate disregard for the integrity of the information, as we will examine in more detail below.

The prime example of the instrumental-strategic use of disinformation for profit would be the entrepreneurial type. For the entrepreneurial type primarily interested in profit, the efficiency and effectiveness of a product are not a priority. For entrepreneurs, the key determinant is the product's market performance. We can also immediately find this type in the homeopathic community. For example, Heel, one

---

[1] For more details, see the clip available at https://medien.bph-online.de/7-bph-film-mit-der-homoeopathie-geht-neue-welt-auf/. Accessed 5 Dec 2023.

[2] For more information on the claimed autism treatments, see e.g. *Librate* at https://www.lybrate.com/topic/homeopathic-remedies-for autism/0778f50c492f6aea6d66da37fc2e48b. Accessed 2 Dec 2023.

[3] For more information on the claimed cancer treatments, see e.g. *Memorial Sloan Kettering Cancer Center* at https://www.mskcc.org/cancer-care/integrative-medicine/herbs/homeopathy. Accessed 2 Dec 2023.

[4] For example, consider the *'10:23' Campaign*, where critics and skeptics of homeopathy consume an entire packet of highly potentized homeopathic globules to demonstrate its ineffectiveness first-hand. From a classical homeopathic perspective, consuming an entire packet should result in an overdose or at least elicit some reaction. Yet, no reaction occurs.

of the world's largest manufacturers of homeopathic remedies, claims on their packaging and in the patient information that their products have a specific healing effect despite the lack of scientific proof. Neurexan, for example, is said to "help against nervous restlessness and sleep disorders and lowers cortisol levels during stress."[5] Entrepreneurs can use targeted disinformation to influence potential buyers' purchasing decisions.

In this paper, we take a broader look at the phenomenon of disinformation in the context of homeopathy. To better understand disinformation's effectiveness and dissemination potential, we propose not only considering the entrepreneurial type, as we have described, as a potential producer of disinformation but also closely examining the collective of homeopathic followers whose beliefs and attitudes shape the information they produce and share.[6] The ideal-typical categorizations of homeopathic publications developed in the following should help to gain a more differentiated view of the broad spectrum of attitudes of the narrative carriers.[7] Our claim to be shown is that the attitudes and opinions of the homeopathic narrative carriers are constitutive in the reception and dissemination of (dis)information. As we will show, narrative carriers are not always seriously epistemically convinced; in some cases, we also find a latently ironic approach to homeopathy among narrative carriers, in which a certain awareness of the falsity of the information resonates.

2 Typology

In the following, we present an outline for a typology of homeopathic narrative carriers, which we use to illustrate the different attitudes and approaches to homeopathy.[8] For this approach, we propose to consider not only the degree of conviction that motivates someone to use homeopathic remedies but also the extent to which this conviction is expressed practically – be it in a family setting, in a speech, a post on social media (tweets, etc.) or practical action. While our first approach enables a detailed description of epistemic attitudes – which can translate into consumer decisions – our second approach allows a more differentiated view of the performative dimension of those beliefs, in which the different epistemic attitudes become visible.

---

[5] For more information on the claimed effects, see *Neurexan Patient Information* at https://neurexan.de/neurexan-service. Accessed 2 Oct 2023.

[6] Thus, the concept of disinformation, as it is defined at the homepage of the Bundesministerium für Bildung und Forschung (BmBF), is too narrowly focused on the strategic aspects: https://www.bmbf.de/bmbf/de/home/_documents/forschung-gegen-desinformation-im-internet.html. Accessed 4 Oct 2023.

[7] In the following description, we understand narrative carriers to be all those who use homeopathic remedies or at least have a positive attitude towards them. Narrative carriers tend to be more receptive to positive reports about homeopathy and also tend to report positively about it.

[8] We are aware that, in reality, there are mixed forms of these types among homeopathic narrative carriers. However, for the following discussion, we are initially interested in a clear conceptual distinction.

With regard to epistemic attitudes, we propose first to separate two different forms: those of the True Believer and those of the Committed. At first glance, both groups seem to belong to the same concept but differ in their internal structure. True believers do not need commitment to uphold their convictions about what they consider right. They rest in the unshakeability of his conviction. In contrast, a commitment presupposes the awareness of a possible error. The committed person recognizes an inherent risk in his conviction. This is why commitment is needed: commitment levels out the risk in a subjective sense, so that the person can speak with a certainty similar to that of a true believer. On the one hand, the True Believer assumes the objectivity of his belief; he acts to the best of his knowledge and belief. The Committed, on the other hand, cannot assume objectivity, which makes a commitment necessary.

This gives us a rough distinction between the True Believer and the Committed: Unlike the True Believer, the Committed acts in the awareness of a possible error. He could also be wrong, but since he has committed himself to his views, he tends to fall back on strategies that make his belief appear true or at least not untrue. In this regard, the Committed will be able to speak about homeopathy in the same way as the True Believer, *and* on the other side, he can make statements that clearly have a strategic character. From an observational perspective, it is difficult to distinguish between the two epistemic attitudes, as the Committed can also talk like a True Believer and be confused with him. This is part of his strategy. However, depending on the example, it is possible to give reasons why a certain example is more likely to be a committed type. The decisive factor here is how much the strategic character of a performance stands out – and this is a matter of interpretation, as we will show in the following three examples.

As a first example, let's look at the clip "Homeopathy Cured Me" from the *Federal Association of Patients for Homeopathy* (BPH).[9] In the clip, an initially skeptical patient reports her homeopathic healing success with "abdominal adhesions". After initial relief, the pain returned the next day but steadily improved with continued homeopathic treatment. Ulf Riker, internist and second chairman of the German Central Association of Homeopathic Physicians, interprets the patient's rapid initial response as partly due to a placebo effect. However, he attributes the continued healing success to the effect of homeopathy. The internist, Riker, and the patient appear as convinced narrative carriers in the clip. Both appear authentic in their performance. The fact that Riker interprets the patient's initially rapid positive reaction as a placebo effect and thus initially rejects the supposed positive effect of the remedy also contributes to his authenticity. He thus demonstrates that he is familiar with the effect of the remedy and knows how to distinguish the true positive effect of the remedy from a supposed positive effect. In this sense, he appears self-critical and confident from an alleged factual point of view.

---

[9] For more details, see: https://www.homoeopathie-online.info/bph-film-die-notoperation-hat-mein-leben-gerettet-die-homoeopathie-hat-mich-geheilt/. Accessed 5 Dec 2023.

However, we must be careful when interpreting the clip: We are unaware of the doctor's exact motives and background. Moreover, it is a clip that the association produced specifically for homeopathic advertising purposes. Authentic persuasion without any commitment is undoubtedly the most effective form of advertising, but we cannot rule out the possibility of a commitment in the background. This is because a committer is most convincing when he covers up his subjective commitment and assumes the objective validity of his statements. Therefore, the apparent conviction could merely be the result of strategic staging. A supposedly convinced person could actually be committed.

The following example is from an interview with Michaela Geiger, the first chairwoman of the German Central Association of Homeopathic Doctors, featured in the *Südwestrundfunk* (SWR) documentary "Homeopathy – Healing or Humbug?".[10] In contrast to Riker's self-produced media content, Geiger faces critical questioning from SWR reporters. The interview puts her in a position where she must address scientific criticism and widespread skepticism about homeopathy. When asked, "What are the drugs actually supposed to do in the patient?", Geiger responds (3:18–3:36): "As a practitioner, I see the effectiveness. How it works, or how the drug reacts in the non-molecular space, is something for further research, a task for basic researchers to uncover the detailed causal mechanisms." Her answer allows several observations to be made. Firstly, her response is notable for its rhetorical nature; she deflects the direct question, emphasizing future research instead of current evidence or an explanation of homeopathy's causal mechanisms.

Secondly, Geiger's response appears to align with what we termed the committed type, as it seems she is uncertain and does not truly believe in homeopathy. However, her answer takes on a different dimension when considering the unique characteristic of homeopathy – its effectiveness claimed by practitioners despite an unexplained causal mechanism. Homeopaths, like Geiger, often express firm belief in their method's efficacy, even though they cannot precisely explain how it functions within the organism.[11] This situation is reminiscent of early views on Penicillin, as Gerring noted (2010, 1505), where belief in its efficacy preceded a full understanding of its mechanisms. It indicates that it's not uncommon to be unsure about the exact causal mechanism before a substance is widely used. Thus, her statement might also fit the true believing category, reflecting a deep conviction in homeopathy's effectiveness despite the lack of clarity about its causal mechanisms.

Let us now conclude with a third, clear example from the pro-homeopathic movie "Magic Pills":

> I have been a homeopath for 20 years, and my husband has been a homeopath for over 30 years. Do you really think that anybody would want to be a homeopath if it didn't work? What do you get from it? You get slander, you get abuse, you get people who are telling you on parties that what you are doing is bullshit. You get absolutely no respect whatsoever. You

---

[10] See the full documentation at SWR Doku "Homöopathie - Heilung oder Humburg?", YouTube, 29 Jan 2020, https://www.youtube.com/watch?v=NrhpKNebaKo. Accessed 15 Nov 2023.

[11] See *Basic Knowledge of Homeopathy* Chap. 2 at https://basiswissen-homoeopathie.de. Accessed 2 Dec 2023.

> always have to justify what you are doing. Why would anybody want this in their life? 20 years of shit. Why would I do it if it didn't work? Why would I be in Africa if I wasn't doing something good? I wouldn't do it if I didn't get great results and if at the end of the day, I didn't know that what I'm doing is really good, really great and makes a difference to people's life. And if I didn't think homeopathy wasn't completely completely amazing. I wouldn't do it.[12]

Here, Camila Sherr, a homeopath, defends her many years in the field and seeks to dissociate herself from allegations of misleading practices or "bullshitting". According to Harry Frankfurt, "bullshit" is characterized by indifference to the truth (Frankfurt, 2006, 40). By this definition, Sherr's defense does not constitute bullshit as she emphasizes her conviction and personal experience, even if the rhetorical intensification such as "completely completely amazing" may detract from the seriousness of her statement. Sherr's emphasis on her authenticity and commitment to homeopathy and her depiction as a victim of critical public voices reflects her cognizance of the potential critiques facing homeopathy. Her acute awareness of the inherent risks associated with her beliefs and her forthright defense of her position typifies her as a committed person. This portrayal suggests a deep engagement with the subject, indicating that her commitment stems from professional involvement and a personal conviction in the efficacy and validity of homeopathic practices.

Using Sher's example, we can deduce that in distinguishing between the true believing and committed type, the true believing type represents a figure that is inherently laden with presuppositions. In the case of people who actively promote the homeopathic narratives and regularly face opposition, it is logical to categorize them as committed. As we have previously discussed, this immunization strategy, where staging certainty is prominent, emerges as a notable and provocative approach. These examples highlight the importance of contextual knowledge in classifying types. It is only when the strategic nature of a speech or performance is exceedingly apparent that we can confidently exclude the possibility of encountering a true believing type.

The True Believer and the Committed person spread positive information – or possibly also disinformation-type information, as we will discuss – about homeopathy because both types are supporters of the homeopathy movement. However, they are not the only ones spreading positive information about homeopathy. As we will now describe, there is also the group of bullshitters and semi-ironic persons who can be considered for dissemination. These types are not, or at least not necessarily, supporters. They have other motives.

Let's start with the bullshitter. This type spreads information about homeopathy affirmatively and often provocatively, not primarily out of a belief in its healing power but rather to position themselves against established institutions such as science, governments, and conventional medicine. In our understanding, a bullshitter might also engage with other topics, like climate change denial or promoting absurd conspiracy theories, as a way to express their mistrust. For example, the Narayana-Verlag claims that "professors of medicine" have "intimidated" the "government

[12] Camila Sherr in *Magic Pills* at https://magicpillsmovie.com/. Accessed 2 Dec 2023.

and health authorities" to suppress homeopathic remedies, which they allege can "truly" cure cancer (Hubbard, 2012, 1). Alternatively, the Zeiten-Schrift endorses homeopathy (Münker, 2020, 10) and raises awareness about the "techno-dictatorship", purportedly arising with the advent of the coronavirus (Bellotto, 2021, 24).

Of course, we could interpret the examples mentioned as instances of committed persons who genuinely believe in homeopathy. We would ultimately be able to verify this empirically in the individual cases. However, our primary aim is to establish clear conceptual distinctions. For example, a bullshitter in the context of the homeopathy discourse might gravitate towards homeopathy primarily as a means to differentiate themselves from established institutions rather than due to a belief in its efficacy. The *Zeiten-Schrift* – a magazine that claims to offer "more quality and truth in life," is known for its favorable stance on homeopathy. It also delves into conspiracy theories. This includes speculations about who was responsible for the assassinations of the "three Kennedys" and assertions about humans being "beings of light". Such narratives are characteristic of what we term bullshit. If we disregard the monetary aspect for a moment, then we can also see this magazine itself as an expression of such bullshitting.

Considering that the bullshitter in the homeopathy discourse is not entirely serious about homeopathy, unlike the True Believer, we could also confuse him with the semi-ironic type. By this type, however, we mean a different figure, which to a certain extent, mirrors the bullshitter: For this type trusts the established institutions, thus, also takes the criticism of conventional medicine towards homeopathy seriously. However, the criticism does not stop this type from taking homeopathic remedies. For example, In the SWR market check, the reporter asks, "Did these remedies also work?" The interviewees answer, "Well, arnica definitely did, even if it was just the sweet taste" (1:52); "It does work! We imagine that it works" (2:07).[13] Here, the conviction about the healing effect itself is strongly relativized. The semi-ironic take the remedy in the knowledge that it doesn't actually work, which lends an ironic note to their homeopathic consumption. In this sense, the semi-ironic type has no commitment or a very weak commitment to homeopathy, similar to the bullshitter, as they do not truly stand up for their belief.

3 Different Types of Disinformation in the Homeopathy Discourse

If we summarize these considerations, we arrive at the following typological overview (Table 1):

[13] See the SWR market check at SWR Marktcheck, "Die Tricks mit Globuli und Co.: Das Geschäft mit der Homöopathie", YouTube, April 23 2020, https://www.youtube.com/watch?v=mZjtx0crJ4w. Accessed 2 Dec 2023.

Table 1 Types of narrative carriers in Homeopathy Discourse

	Epistemic	Performative
Type A: True believers	Convinced	Without contradiction
Type B: Committed persons	Committed	Latently Strategic
Type C: Bullshitters	Not or very weakly committed	Spreading distrust
Type D: Semi-ironic persons	Not or very weakly committed	Ironic non-committal consumption
Type E: Entrepreneurs	Indifferent	Sales

The typology outlined here helps us to analyze different forms of disinformation in homeopathy discourse in more detail. Our analysis is based on the consideration that the different types of A–E tend to share and disseminate different forms of information about homeopathy, some of which can be characterized in more detail as disinformation.

We propose to characterize information as disinformation[14] if the information disseminated is false information that is either intended to deceive a person (Modality I) or to articulate mistrust of established institutions such as the mainstream media or 'science' (Modality II). We, therefore, differentiate the modality of disinformation according to the underlying intention of the false information. In the disinformation of the first modality, the strategic aspect is decisive. In contrast, the disinformation of the second modality is the expressive content of the disinformation, as we will explain in more detail below using examples from the homeopathy discourse (Fig. 1).

As far as disinformation of Modality I is concerned, we can go back to the example of the company (type E) mentioned at the beginning, which deliberately spreads false information in favor of its product (Neurexan "lowers cortisol levels during stress") to encourage its customers to buy it. This type of disinformation is based on the fact that the potential customer considers the source or the company responsible for the source trustworthy. Therefore, the Entrepreneur's act of providing information implies a breach of trust, as the customer's trust is exploited.

In disinformation of Modality II, we can now consider a bullshitter (type C) who spreads deceptive information about homeopathy, not because he is primarily concerned with homeopathy, but because he wants to agitate other people. Homeopathy is a proxy for positions that are directed against the mainstream (Goldenberg, 2021, 91). As an example, we cited an entry from the blog of the Narayana publishing

[14] A very precise analysis of the term *fake news*, which we understand here as a synonym for disinformation, can be found in Jaster & Lanius, (2021, 23). In their paper, they provide an overview of analyses of the *fake news* term by other authors. A common aspect of these analyses of the fake news concept is the attempt to work out the basic characteristics of fake news or disinformation in general. In contrast, as we emphasized in the introduction, we concentrate on the motives of the disseminators of disinformation and, based on these motives, distinguish between different types of disinformation. We are therefore less concerned with the question of what disinformation is and how it can be defined, but rather what forms it can take depending on the motives of its disseminators.

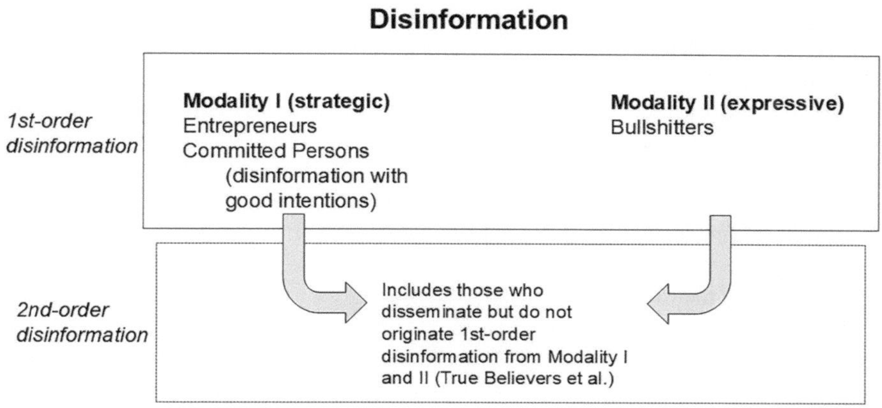

Fig. 1 Types and motives of disinformation

house, according to which "medical professors" had allegedly intimidated "government and health authorities" to prevent homeopathic remedies. It is about articulating mistrust. There may well be other people who believe such statements and consequently believe in the healing power of homeopathic remedies to fight cancer – but this effect is secondary. Unlike Modality I disinformation, Modality II disinformation is not primarily about spreading a particular belief but about articulating mistrust. In this case, false information is used mainly because of its expressive function, which lies in its agitative power.

We have thus clarified the extent to which the two modalities of disinformation differ. These are two fundamentally different modalities, although the effect can be similar. We now want to examine additional cases that further differentiate Modality I of disinformation. To do this, let's look at the remaining cases from the typology above.

Let us first take type A, the True Believer. Since we assume that this type of person is entirely convinced of the healing effect of homeopathy, it cannot be a case of disinformation. Suppose a person of this type informs another person that homeopathy can successfully cure cancer. In that case, they are spreading false information, but according to our proposed definition, this is not a case of disinformation. However, it may be a case of second-order disinformation, for example, if the person in question has taken the information from a source that is deliberately spreading disinformation.

Now we turn to type B, which are the Committed Persons. The Committed, as we have characterized, are convinced of their cause but do not necessarily believe that they have the arguments on their side. When their convictions are critically questioned, they engage in strategic behavior to defend them. They resort to rhetorical means to prove themselves right. One rhetorical device can be to refer to supposedly scientific studies, which in turn are deliberately disinformation. This act of referring to disinformative studies is then a case of disinformation with good intentions, as the person is deceiving in the interests of a cause they advocate for

Finally, we are left with type D, which we have classified as semi-ironic. A person of type D takes homeopathic remedies but without being seriously convinced of their effect, as they trust that the criticism of homeopathy from the scientific community is correct. Since type D strongly relativizes the conviction towards homeopathic remedies, we assume this type lacks the motivation to share information about homeopathy that could be characterized as disinformation. There is no will to spread false information for a specific purpose of deception, nor is there a mistrust of established institutions such as science, which could find expression in disseminating false information about homeopathy. In this sense, this type does not produce another kind of disinformation worth mentioning.

The various forms of disinformation, as can be summarized, depend on the motives of the homeopathic narrative carriers. As mentioned in the introduction, this paper aimed to clearly articulate the conceptual differences between the various types of narrative carriers. Therefore, we have only discussed cases where a specific motive is crucial in the dissemination of disinformation. To conclude this part of the discussion, we would now like to address the relationship between the various motives, as not all are mutually exclusive. For instance, the motive of the Entrepreneur (Type E) can be combined with the motives of all other types (A–D). In this sense, someone might spread disinformation because they want to articulate their distrust of institutions (Type C). At the same time, this way of using disinformation expressively can also be a monetary business model, as it generates attention. The *Zeiten-Schrift*, which recommends not only homeopathy but also spreads conspiracy narratives, could be cited as an example of this. It is also possible, for example, that a person in the sense of Type A is interested in selling homeopathic remedies. Then, this person, as a seller, believes in the homeopathic product in an authentic sense but also sells it out of a monetary interest.

4 Conclusion

In our paper, we looked at the different types of homeopathy narrative carriers (True Believers, Committed persons, Bullshitters, Semi-Ironic persons, and Entrepreneurs). We analyzed the specific forms of disinformation they spread. To do this, we first looked at the motives of the narrative carriers. We argue that the type of disinformation depends on the motives.

We made a fundamental distinction between a strategic and an expressive modality of disinformation. In the first modality, under which we subsumed the Committed and the Entrepreneur, false information is spread for deception. To convince people of a certain cause, the Committed persons use disinformation with good intentions to assert and maintain their position in the discourse. The Entrepreneur uses it to encourage customers to buy a product. In the second modality, essentially the bullshitter, false information is spread to articulate mistrust of established institutions (science, governments, conventional medicine, etc.).

We categorized the semi-ironic and the true believing type as non-disinforming. True Believers may spread false information, but they do so out of deep conviction and not with the intention to deceive. The True Believer, therefore, does not act out of a disinformation motive unless he refers to a deliberately disinforming source (second-order disinformation). The Semi-Ironic lacks a firm conviction of the effectiveness of homeopathy. They are also not keen to share false information and show no mistrust of established institutions. Instead, he questions scientific findings and is quick to relativize his consumption, so we do not assume that significant disinformation arises here.

In the framework of our typology, it has become clear how difficult it is to distinguish between the types and how many borderline cases emerge on closer examination. The framework helped us to provide a supplementary perspective on the phenomenon of disinformation. Depending on the interpretation, the supposed disinformer turns out to be someone who acts with full conviction and merely spreads false information.

At a time when even a former US president has accused established media of spreading "fake news",[15] and in view of the constantly changing media landscape, which is characterized by the emergence of social media and the sheer flood of information, it is precisely this differentiation in the understanding of disinformation that seems to us to be of great importance. Without a sophisticated understanding of disinformation, we cannot even begin discussing countermeasures. Our proposed approach to categorizing disinformation is not only applicable to the field of homeopathy but can be applied wherever narratives are supported by more than just entrepreneurial motives.

References

Bellotto, S. (2021). The great reset: Die welt nach corona. *Zeiten-Schrift, 104*, 24–41.

Frankfurt, H. (2006). *Bullshit*. Trans. Michael Bischoff. Suhrkamp Verlag.

Gebhardt, K.-H. (1991). Beweisbare Homöopathie. In G. Eberlein (Ed.), *Schulwissenschaft, Parawissenschaft, Pseudowissenschaft* (pp. 71–82). Hirzel, Wissenschaftliche Verlagsgesellschaft.

Gerring, J. (2010). Causal mechanisms: Yes, but…. *Comparative Political Studies, 43*, 1499–1526.

Goldenberg, M. (2021). *Vaccine hesitancy: Public trust, expertise, and the war on science*. University of Pittsburgh Press.

Hubbard, Bryan. 2012. Viel mehr als ein Placebo: Homöopathie drängt den Krebs zurück. Retrieved December 10, 2023, from https://www.narayana-verlag.de/spektrum-homoeopathie/viel-mehr-als-ein-placebo-homoeopathie-draengt-den-krebs-zurueck.

Jaster, R., & Lanius, D. (2021). Speaking of fake news: Definitions and dimensions. In S. Bernecker, A. K. Flowerree, & T. Grundmann (Eds.), *The epistemology of fake news* (pp. 19–45). Oxford University Press.

[15] See Sky News, "I'd like to welcome the fake news media", YouTube, Nov 9 2022, https://www.youtube.com/watch?v=g_rza9DCCi8. Accessed 2 Dec 2023.

Jütte, R. (1996). Wo alles anfing: Deutschland. In M. Dingens (Ed.), *Weltgeschichte der Homöopathie. Länder – Schulen – Heilkundige* (pp. 19–47). C.H.Beck.

Lübbers, C., & Endruscheit, U. (2021). Homöopathie – eine Therapieoption für die Praxis? Bewertung unter dem Blickwinkel der evidenzbasierten Medizin. *HNO, 69*, 679–690.

Mukerji, N., & Ernst, E. (2022). Why homeopathy is pseudoscience. *Synthese, 200*, 394.

Münker, B. (2020). Corona: Mit Homöopathie die Pandemie bekämpfen. *Zeiten-Schrift, 103*, 10–11.

How to Lie Without Lying: Exploring Strategies of Assertion-Denial

Jörn Wiengarn

Contents

Abstract In our (presumable) post-truth era, the constant and public assertion of falsehoods isn't the only concern; equally alarming is the fact that the disseminators of disinformation seem quite successful at evading responsibility when held accountable for their lies. As I argue, they achieve this by employing one of two basic strategic options. First, they simply refute the falseness of their claims. Second, and this is an often-overlooked strategy, they deny that their statement was ever intended as a truth claim at all. More precisely, the strategy consists in portraying a previous utterance as a non-assertive speech act, thus denying any epistemic responsibility for it. This is what will be called *a strategy of assertion-denial.* The aim of this article is to explore the working mechanism of this strategy and to point out its appeal to those with an intention to deceive. Depending on the respective re-interpretation of one's previous speech act, I will furthermore distinguish and explain three key sub-variants of assertion-denial: the *"It's just my opinion"*-, the *"It's not me who claimed this"*- and the *"That was meant sarcastically"*-variant.

J. Wiengarn (✉)
TU Darmstadt, Darmstadt, Germany
e-mail: joern.wiengarn@tu-darmstadt.de

M. M. Resch et al. (eds.), *Trust and Disinformation*,
https://doi.org/10.1007/978-3-031-96790-0_4

1 Introduction

There is something puzzling about the post-truth era we live in: We are undeniably confronted with a staggering number of fakes and often blatant lies nowadays. Yet only rarely do we see people who admit to having lied in public. How is this possible?

Here are two possible explanations for this conundrum: First, maybe those spreading falsehoods and lies often escape immediate fact-checking, thereby evading responsibility and remaining undetected or at least unchallenged with their lies. Second, even when held accountable, such individuals may go into an epistemic defense mode, disputing the falsehood of their statements, and thus potentially managing to uphold the lie. Call this the *epistemic-rebuttal strategy*.

This article is about a third, less explored possibility, another kind of strategy that liars use in order to avoid being held accountable. At its core, the strategy involves retroactively denying that what one said was ever intended as a serious claim about facts, allowing the speaker to protest: "I did not lie. In fact, I did not even make a proper assertion at all". I will refer to this as the *strategy of assertion-denial*. In more technical, speech-theoretical terms, one can describe its working mechanism as follows: The strategy does not mount a *defense on the truthfulness of a statement* but aims to give its *illocutionary* dimension a specific twist, thus aiming to present it as *categorically* exempt from classification as a lie. More specifically, the speaker attempts to reclassify their statement as non-assertive, thus as not including a truth claim. To put it more bluntly, the strategy works as follows: When called out for a lie, a liar can slip off the hook by insisting: "I didn't mean it *that way*".

On a closer look, this strategy appears not only pervasive. It also seems to have a whole plethora of variations, which range from claiming that one was just expressing one's opinion, to insisting that one did just pass on what another source claimed, to claiming that one was just being sarcastic. To explore some of these sub-variants is the aim of this article. As I will argue, what they all have in common is that they allow a speaker to retroactively deny or diffuse their epistemic responsibility by denying that they were making a proper assertion at all. In doing so, They provide a means to get a message out, while allowing the speaker to eschew, or at least muddle, genuine commitment to its truth. What they differ in is the particular interpretation they offer of the prior speech act.

Here is how the article is structured. I will first give an account of lying that underscores its assertive character and thus the normative responsibility that a liar both assumes and fails to live up to. Based on this, I will second elucidate how strategies of assertion-denial function and what kind of benefits they offer from the perspective of a person with an interest to deceive. Lastly, I will delve deeper into three variations of this strategy. They will be called: the *"It's just my opinion"*-, *"It's not me who claimed this"*- and *"That was meant sarcastically"-variant*.

However, before we delve into this, a couple of remarks seem to be in order. First, although we will primarily be concerned with lying and a particular strategy individual liars can use to keep their countenance, it is important to note the strategy's wider application range. Thus, it can just as well be used by a mere bullshitter, i.e., by someone who does not, like a liar, exhibit an intention to deceive, but rather a general lack of concern for the truth. Yet, our emphasis remains on lying as a paradigm case to work out the logic of the strategy of assertion-denial. That is because lying is, as will be shown, by definition based on (or rather exploiting) the normative commitment of assertive speech acts, while this must not necessarily be the case for an instance of bullshitting.[1] The strategy of assertion-denial is therefore naturally suited for a liar, but not necessarily for a bullshitter.

Second, the strategy of assertion-denial will be presented as a strategy that can be used when a liar is called out. It is in such scenarios that a liar may retroactively reinterpret their previous false statement. For simplicity's sake, I will restrict myself to treating this particular situation. Nonetheless, the logic behind the strategy can clearly also be used *in anticipation*: A crafty liar can make preparations to be able to use this strategy most effectively should they get into trouble. As we will see below, the strategy taps into the fact that norms about what counts as a serious assertion leave some principled room for interpretation. A cunning liar can consciously push things further in their favor, by expressively ambiguating, obfuscating or leaving open how their utterance can be understood. By doing so, they can make the strategy of assertion-denial work most effectively. Relatedly, long term strategies of disinformation campaigns could aim at muddling the standards of public speech to facilitate the future use of assertion-denial. To not exceed the scope of this essay, however, we will have to leave aside these anticipatory strategies and rather focus on assertion-denial as such.

Third, note that the list of sub-variations below should be seen as a first proposal. I do not claim that the list is exhaustive. They represent some of the most notable and intriguing variations. But it's quite likely that other nuanced tactics exist.

Lastly, note that I will not discuss any political-ideological affiliations of the strategies below. Of course, one could argue that some of them resonate with certain ideological or political beliefs: For instance, the first and third variation I'll discuss below seem to resonate with a libertarian ideology that highly values individual freedoms, such as freedom of opinion and expression (while often abstracting from responsibilities accompanying such freedoms). The second variation seems prime for a populist rhetoric, where speakers portray themselves as mere representatives or mouthpieces for a broader group (typically referred to as "the people"). In principle, however, I think that all the sub-strategies discussed below can universally be used to disseminate *all kinds* of falsehoods and therefore serve *any possible* political agenda or ideology.

[1]At least in Frankfurt's (2005) original account, bullshitting is not explicitly tied to assertive speech-acts. However, considering that bullshitting, as Frankfurt describes it, often involves the use of vague language, it can serve as an ideal precursor to effectively employing the strategy of assertion-denial.

2 Lying and the Normative Commitment of Assertions

Before delving into the strategy of assertion-denial, we need to get a clear handle on the notion of lying and its assertive nature. According to a standard view, lying is defined as making "a believed-false statement to another person with the intention that the other person believe that statement to be true" (Mahon, 2016). However, as Chisholm & Feehan (1977) have argued, lying requires more than just the making of a statement, but more specifically the making of an *assertion* or *telling* someone something.[2] By their view, a liar isn't just someone who utters a falsehood, but someone who openly *asserts* something while knowing that it is false.

While this assertion-component is not universally accepted as a necessary component of the definition of lying,[3] I will assume it for present purposes. At any rate, even if one can assume a broader definition of lying, my focus will remain on cases, where the liar *tells* a lie, i.e., where they make an overt assertion about something they know is false.

The natural follow-up question, which will now be given more treatment, is of course: What exactly is an assertion or an act of telling? Like the topic of lying, this is in itself a broad question which, not just since the emergence of speech-act theory, has generated extensive philosophical debate. However, for the purposes of the present discussion, I'll hone in on its normatively binding force: the fact that a speaker commits themselves to the truthfulness of what they say.

In his article "Getting Told and Being Believed", Richard Moran argues that telling someone that p amounts to giving an assurance that p is true. This is what he calls the Assurance View of testimony. He contrasts this view with the Evidential View of testimony, according to which, telling someone that p merely consists of delivering a piece of evidence for p. Although this view has been prevalent within the philosophical tradition, it faces a major difficulty, namely that it cannot account for what it actually means to *tell someone something*. As Moran argues by drawing on Angus Ross:

> Speaking is not like allowing someone to see you are blushing. The problem is not, however, that the fact of our having chosen to use certain words cannot be evidence for some further conclusion. Our choices can certainly be revealing. The difficulty lies in supposing that the speaker himself sees his choice of words in this light, which in turn makes it difficult to suppose that this is how the hearer is intended to see his choice." (as cited in Moran, 2005, 8).

Deliberately letting someone know that p by showing them evidence for p is something categorically different than telling someone that p. An initial difference between the two lies in the fact that a person's freedom plays a different role. Evidence is supposed to speak for itself. The fact that *I deliberately* show you some evidence (for example a photography of a dead body in the park) is incidental to the

---

[2] Note that for present purposes, I will not distinguish between telling and making an assertion like Faulkner (2013) proposes.

[3] For an overview, see: Mahon, 2016.

evidential force of that evidence. Things are different, however, if I were to *tell* you that there is dead body in the park. Here, my deliberate intention to do so is essential. That is why if I were to murmur in my sleep or in some other sub-conscious state that there is a dead body in the park, this would not count as an act of telling.

As Moran argues, freedom is constitutive for an assertive speech act with which a speaker assumes responsibility for the truth of what they say. This idea is central to what he calls the Assurance View of testimony. According to this view, a teller is not just presenting evidence for what they say, but through their assertive speech act, they give an assurance. They thereby assume responsibility for the truth of what they say which alters the normative relationship to the hearer.[4] It is for this reason that a teller can be held responsible, which implies that they can be pressed for justification for the truth of what they say, or that they can be blamed if what they say turns out to be false, let alone if they told blatant lies. As Timothy Williamson puts it: "To make an assertion is to confer responsibility (on oneself) for the truth of its content; to satisfy the rule of assertion, by having the requisite knowledge, is to discharge that responsibility, by epistemically ensuring the truth of the content" (Williamson, 2002, 268 f.).

Additionally, an assertive speech act can be characterized not only by the responsibility the speaker assumes in making it, but also by the speaker's intention. This is another way of flagging the difference between making an assertion and delivering evidence: Through an assertion, a speaker aims at being fully believed and not just at his assertion being considered. This thought is illustrated nicely with an example from Aron Keren: imagine Sandy who asserts to us: "Take my word for it, p". Now.

> suppose that we form the belief that p, and that we do so without weighing other relevant evidence available to us. It seems that Sandy cannot now turn to us and say: "You are correct. But you should have considered all the evidence available to you before reaching that conclusion." […] If she were to criticize us in this way, she would seem to be withdrawing her original appeal to us that we take her word for it. Now compare this with the case of Evelyn, who intentionally provides us with very good evidence for p […]. Suppose that here we again form the belief that p, without weighing other available evidence. If Evelyn's only intention was to provide us with very good evidence for p, then it seems that she can criticize us by saying: "You are correct. But you should have considered all the available evidence before reaching that conclusion." (Keren, 2014, 2600).

In this scenario, Sandy can be seen as merely delivering evidence, while Evelyn makes a proper assertion. Consequently, unlike Sandy, Evelyn can be taken to aim at just being believed. This apparently implies that she expects from the hearer to accept her statement without considering additional evidence that might support or

[4] It can be debated whether a teller assumes primarily responsibility for the *truth* of what they say as Moran seems to assume. In contrast to Moran, Hawley (2019, Chap. 3) argues that a teller commits to being *truthful*, which implies that they can fully live up to their responsibility, even if their claim turns out to be false, provided that they made the claim with the best and most sincere intentions. For present purposes, however, the difference is not of major importance. What matters is that there is some kind of truth-related normative commitment on the part of the teller. Whether or not a teller commits to the truth or to being truthful, the logic of the kind of strategies I will discuss below remains the same.

contradict her claim. Moran hints at the same point when he gives an alternative formulation of the difference between telling and delivering evidence: A speaker does not view his own assertion, "as if the meaning of his utterance were 'Now I have spoken; make of it what you will' rather than 'Take it from me'" (Moran, 2005, 26). Another way of putting this idea comes from Edward Hinchman (2014), who holds that with an assertive speech act (claiming that p) a speaker seeks to conclude an addressee's deliberation whether p. Using Hinchman's terminology, we can refer to this feature of assertions as their *"closure conducive"-character*. This perspective also sheds new interesting light on the exact contours of a speaker's responsibility: Thus, an assertive speech act does not merely commis the speaker to truthfulness, making it possible that they can be asked for reasons supporting their clais. It also commits to a *higher-order* guarantee that the question whether p can be regarded as *settled* and not requiring any *further* inquiry.[5]

No matter how we delineate the exact range of a speaker's accountability, they only assume such responsibilities insofar as they publicly perform an assertion. Yet here also lies a practical difficulty: conventional norms defining what counts as a committing assertion *always allow for some degree of interpretation*. Under normal circumstances, we roughly share a common understanding of such conventions discerning when an utterance counts as an act of telling and when not. For instance, if I raise my voice to adopt an ironic tone while talking clear nonsense, it's commonly understood not to be meant as a serious claim. In contrast, statements made under oath in a courtroom are paradigm examples of normatively binding assertions for us. However, many utterances fall into a gray area, making it ambiguous whether they represent claims to truth. We may know this from fine-crafted satire that deliberately plays with this ambiguity. But as I aim to show in the following, it is precisely this leeway of conventional norms that the strategy of assertion-denial I will be addressing now can tap into.

3 What a Liar Can Do: The Strategy of Assertion-Denial and Its Benefits

Just for a moment, imagine yourself in the position of a liar with a strong interest in deceiving others or leading the public astray: From this perspective, the norm to be truthful, the commitment to the truth one makes in asserting is quite a nuisance for you since you can be held accountable by it. However, there are several ways for you to react to this difficulty. For one, you could of course just bite the bullet and admit (possibly with a shrug of the shoulders) that you have lied when accused. While this option might offer a brief moral redemption, it comes with the obvious

[5] As Hinchman (2014) stresses, whether it should be seen as settled depends on the specific needs of the addressee. Depending on the risks at play for the addressee, a speaker should fulfill higher or lower doxastic standards in order to warrant that p can be seen as settled.

cost of not being believed. And potentially, you might even risk not being believed in the future ("once a liar, always a liar!"), thus making it impossible to build up a long-term strategy to deceive the public.

A second option could therefore be to go into epistemic rebuttal. For example, you could challenge any evidence against you or make up some alleged counterevidence in your favor. Similarly, you could of course also anticipate such troubles and therefore try to make the lie so slick and professional that it promises to go undetected (or at least for as long as possible). However, this strategy also has its drawbacks, for it is probably no easy feat. It takes an effort to create a convincing and undetectable lie or to defend one against counterevidence. Of course, depending on the vigilance of those fact-checking your claims and calling you out as a liar, you may find an easy escape: Just make up some new alleged facts to bolster your lie, raise some bullocks suspicion about the opposing evidence; and maybe you might just baffle everyone and bullshit your way out of the misery. Yet, this is a gamble. There still remains the risk that the fact-checkers are thorough and counter with overwhelming proof – and then you would really risk losing your face.

However, there are two more strategies you can use. They both consist in retroactively denying that your previous statement included the kind of truth claim that the fact-checkers try to make you accountable for. The first operates on the level of the content of your previous statement. It consists of denying that what you are being accused of is not *what* you actually said. This strategy works most convincingly by employing ambiguous language in your statements, for example by using double entendres (working similar to "dog-whistling"), or by insinuating falsehoods without literally stating them.[6] Yet, while this strategy may be quite "elegant", it also has its limits. You can only use to the extent that you find suitable ambiguous terms or ways to implicitly communicate falsehoods, but this might not always be possible.

At this point, the strategies of assertion-denial come into play. These basically offer you various get-out clauses that enable you to sclaim that what you said was never meant as a proper assertion *at all*, and thus cannot even qualify as a potential lie. They thus allow a liar to play innocent by insisting that they did not assume the kind of responsibility associated with making an assertion, as demonstrated in the previous section. In contrast to the options above, this strategy has the advantage for the liar that it allows them to avoid admitting being responsible while at the same time, first, being relatively effortless and, second, being broadly applicable. With this strategy, there's no need to go into the complexities that a well-crafted epistemic rebuttal might require. It is for this reason that such a strategy can particularly come in handy when the liar realizes they may be running out of arguments when experiencing some push-back. Furthermore, there's no need to craft cleverly ambiguous statements. Strategies of assertion-denial are quite broadly applicable; it can be used regardless of the message you want to get out and is not contingent on finding suitable ambiguous terms or cleverly crafted

---

[6] This is a strategy often used in fake news (cf.J Laster & Lanius, 2021). For an illustration of how a similar mechanism works in propaganda, see: Chap. 4.

innuendos. In essence, assertion-denial allows for a swift and broadly applicable defense mechanism for all those aiming to dodge responsibility when their deception faces scrutiny.

The broad applicability of this concept stems from the principled room for interpretation we mentioned above. Whether someone's utterance can be taken as an assertion can be more or less clear, depending on the specifics of the circumstances. Consequently, it can be more or less clear whether a speaker commits to the truth of what they say. It is this principled ambiguity that the strategy of assertion-denial aims to tap into. It enables liars to articulate a specific interpretation that serves their interests. This is not to suggest that every interpretation is equally plausible; rather this strategy is a universal possibility. How convincingly it works in the end, depends on the original ambiguity of the speech act and the plausibility of the interpretation advanced by the speaker.

So, denying that a previous speech act was meant as an assertive one is the core mechanism of this strategy. Based on this, we can now go on to differentiate different sub-variants of it, depending on the exact (re-)interpretation of the previous speech act they offer. In the following segments, three different versions will be explored.

4 Variations of the Strategy

4.1 "It's Just My Opinion"-*Variant*

The first sub-variant involves the use of a popular get-out clause, namely: "It's just my opinion". Of course, the exact phrasing may vary with alternatives like "I think we just disagree on this". But the precise wording is not what matters here. Let's consider an example of this strategy in use: In a discussion with a Trump supporter, who thinks that Barack Obama is not a U.S.-Citizen, Jordan Klepper from the satirical Daily Show asks him, what kind of proof he has for his view. In response the interviewee claims "I don't have any. It's my opinion" (The Daily Show, 2020).

What exactly is going on in this exchange? Apparently, by stressing that he has just expressed an opinion, the interviewee seeks to cast his utterance in a specific light. Thereby, he apparently assumes that he can be absolved of any responsibility to speak truthfully and, consequently, of any obligation to deliver arguments or evidence for his view. The underlying assumption appears to be that opinions lie outside the realm of truth claims, thus exempting the speaker from any evidential responsibility and potentially allowing him to discard any epistemic burden.

Another way of interpreting what is going on in this example is that the interviewee tries to flag his utterance not as assertive speech act, but as a purely *expressive* speech act. So while Moran is right to note that "Telling someone something is not simply giving expression to what's on your mind" (Moran, 2005, 8); the issue is what *counts* as an act and what as a merely expressive speech act. In the example, the interviewee claims he just meant to *express* his opinion, and nothing more. The qualification 'and nothing more' is important here because, as the theory of the four sides of a communicative act highlights (von Thun & Friedemann, 1981), every speech act contains an expressive *element*. Whatever we say, it always reveals something about us as speakers. However, this does not mean that every statement is *purely* expressive (as when one shouts "ouch!" after hitting one's head at the kitchen table). Yet, the interviewee's strategy seems to require exactly this: to claim that his previous assertion is expressive *through and through*. It is therefore a perfect example of assertion-denial in practice.

Incidentally, an interesting slight variation of the central get-out phrase is "I am entitled to my opinion". This phrase furthermore stresses that one has a *right* as a speaker to express one's opinion. Such a counter yields two additional strong points for someone who tries to evade epistemic responsibility: First, it claims something which is indeed true, the speaker does have a right to express themselves freely (at least within certain constraints). Second, it can help them to portray demands for justification as infringements upon that freedom. Yet, even if the speaker has such a right, as such this simply does not negate their *simultaneous responsibility* as a speaker. And demanding justification as such does not undermine their right to express themselves freely but is a call for responsibility.

4.2 "It's Not Me Who Claimed this"-*Variant*

A paradigm example of this second strategy is the practice of re-posting. As Regina Rini (2017) has argued, re-posting represents a "bent kind of testimony", since if someone shares a news story, it is often unclear whether this is meant as an endorsement of its content. The norms governing how reposts are to be understood are somewhat unstable: it could be that by sharing a post someone wants to encourage a discussion about it, or to just hold it up for ridicule. It is this ambiguity surrounding the act of sharing that provides a convenient defense for social media users who disseminate misinformation.

As a perfect example of how this strategy works, Rini mentions an exchange between Donald Trump and Fox News host Bill O'Reilly. When O'Reilly challenged Trump over a post on Twitter full of fake statistics (for instance, the claim that 81% of white homicide victims are killed by African-Americans, when in fact the number is closer to 15%), Trump replied with a familiar defense scheme: "Bill, am I gonna check every statistic? All it was is a retweet. It wasn't from me" (as cited in Rini, 2017).

What is happening here? Apparently, O'Reilly initially treats Trump's post like an assertion tied to a truth claim about what it says. This is precisely what Trump denies, trying to frame his post as not being meant in an assertive way, but merely as a way of putting information out there. Hence, it is another form of assertion-denial.

The fact that our norms for reposting are not fully settled upon makes social media a perfect environment for such non-committal messaging and the thriving of liars and bullshitters. It plays in their favor by allowing them to get a message out while being able to explain it away when held responsible for it. Yet, this strategy is not confined to social media but has a broader application range. It can have analogues beyond social media and reposting, whenever a speaker claims to have merely passed on a message, portraying themselves as a non-interfering mouthpiece.[7] Symptomatic of this is Trump's habitual use of qualifiers such as "a lot of people tell me that p" or "many people think that p", intentionally clouding whether he actually embraces p. This leaves him some wiggle room to later dissociate from the statement, should it come in handy. As Jenna Johnson from *The Washington Post* characterizes Trump's strategy here: "If the comment turns out to be popular, Trump will often drop the distancing qualifier — 'people think' or 'some say'. If the opposite happens, Trump can claim that he never said the thing he is accused of saying, equating it to retweeting someone else's thoughts on Twitter"(Johnson, 2016).

Note that this strategy naturally connects with the familiar catchphrase that "the audience can make up their own minds". In an NBC town hall meeting, Savannah Guthrie confronted Trump about yet another of his retweets of a conspiracy theory claiming "that Joe Biden orchestrated to have SEAL Team Six, the Navy SEAL Team Six, killed to cover up the fake death of Bin Laden".[8] In response, Trump not only emphasized that it was merely a retweet but adds: "I'll put it out there. People can decide for themselves. I don't take a position." This move, too, is made possible by denying the assertive character of a previous communicative act. Reall that we earlier highlighted the closure-conducive character of an assertion. By denying that his post was meant as an assertion, Trump also denies its closure-conducive feature. This allows him to treat it as if it were merely the transmission of evidence. In other words, he relocates his post from the realm of Moran's "Take it from me" to that of "Now I have spoken; make of it what you will" (Moran, 2005, 26). In doing so, he seeks to shift the entire epistemic burden onto the audience.[9][10]

---

[7] Here, one can see clearly how this strategy is prime for a populist rhetoric, which portrays the speaker as just spelling out what the people think or say.

[8] For the transcript of the town hall event, see: https://www.rev.com/blog/transcripts/donald-trump-nbc-town-hall-transcript-october-15

[9] This tactic likely offers additional advantages for Trump. Most notably, it allows Trump to appeal to a *prima facie* value that holds significant prominence in modern democratic societies: the individual's epistemic autonomy. At the same time, he could depict his critics as patronizing. (However, for a critical discussion of this value, see Zagzebski (2012, Chap. 1)).

[10] Trump's move therefore also stands in stark contrast to what Benjamin McMyler (McMyler, 2011, Chaps. 2.2, 2.3) calls the buck-passing-feature of testimony. According to McMyler, and in line with the warranting character of assertions, if an individual believes that p based on their trust placed in a speaker S who told them that p, the individual can refer back to S when challenged on whether p. In contrast, Trump suggests in the example that the audience shoulders the complete epistemic burden.

4.3 "That was Meant Sarcastically"-*Variant*

The final strategy is another one repeatedly employed by Donald Trump.[11] For example, during a press briefing in April 2020 Trump infamously speculated about the possibility of injecting disinfectant as a potential treatment for COVID-19. When later questioned about the remark, he avoided admitting to his mistake by claiming that what he said was meant sarcastically (Rupar, 2020).

The use of sarcasm or irony is another form of speech that resembles making serious assertions. In fact, irony essentially consists of *mimicking* the act of making serious statements. That is why, when talking ironically, we usually rely on subtle cues to signal our intent, for example changes in tone of voice. In principle though, there is some leeway for interpretation as to what counts as an ironic remark, and it is precisely this ambiguity that can be exploited as a defensive strategy.[12]

5 Conclusion

This article explored a range of strategies that can be used by people who want to get (false) messages out while avoiding accountability for their truth. These strategies all work by denying that one actually assumed the respective responsibility, more precisely by denying that one performed an assertive speech act. All of the different strategies exploit a potential ambiguity about what counts as an assertive speech act. They differ in the kind of alternative interpretation they deliver of the previous speech act. The first strategy does this by claiming that the alleged assertion was a mere expressive speech act, just aiming to share one's opinion. The second one consists in claiming that the previous speech act was just meant to pass on someone else's opinion and aims to portray oneself as a mere mouthpiece. And the third strategy operates by claiming that their previous utterance was meant sarcastically.

All of these tactics thrive where there's ambiguity about what counts as a genuine assertion. Part of broader disinformation campaigns could therefore also be to ensure such ambiguity, be it on the level of individual assertions by crafting them as

[11] It's noteworthy that many examples discussed here involve Donald Trump. In part this can probably be attributed to Trump's prominence and aptness at deploying these strategies. But it is also illuminating to think about the structural reasons such techniques lend themselves specifically to him in his role. Thus, in his role as a (attention-seeking) politician, Trump is regularly held accountable for his lies in public. At the same time, maintaining an image of honesty is crucial for a politician's reputation, making the strategy of assertion-denial particularly useful for him. What seems furthermore important for someone like Trump who aims to appease a diverse range of often conflicting groups, these strategies allow him to convey messages that signal allegiance to certain factions, while affording him the flexibility to backtrack when needed.

[12] Again, some additional advantages of this strategy could be pointed out. For instance, it allows Trump to depict his critics as lacking a sense of humor or as being incapable of grasping sarcasm.

ambiguous as possible, be it on the level of the general norms about what counts as an assertive speech act by intentionally muddling these in the long run. But this is a topic that needs to be explored in a future discussion.

References

Chisholm, R. M., & Feehan, T. D. (1977). The intent to deceive. *Journal of Philosophy, 74*, 143–159.

Faulkner, P. (2013). Lying and deceit. In H. Lafollette (Ed.), *International encyclopedia of ethics* (pp. 3101–3109). Wiley-Blackwell.

Frankfurt, H. G. (2005). *On bullshit*. Princeton University Press.

Hawley, K. (2019). *How to be trustworthy*. Oxford University Press.

Hinchman, E. (2014). Assurance and warrant. *Philosophers' Imprint, 14*, 1–58.

Johnson, J. (2016). *'A lot of people are saying …': How trump spreads conspiracies and innuendoes*. The Washington Post. Retrieved June 19, 2023, from https://www.washingtonpost.com/politics/a-lot-of-people-are-saying-how-trump-spreads-conspiracies-and-innuendo/2016/06/13/b21e59de-317e-11e6-8ff7-7b6c1998b7a0_story.html

Keren, A. (2014). Trust and belief: A preemptive reasons account. *Synthese, 191*(12), 2593–2615.

Jaster, R., & Lanius, D. (2021). Speaking of fake news: definitions and dimensions. In S. Bernecker, A. K. Flowerree, & T. Grundmann (Eds.), *The epistemology of fake news* (pp. 19–45). Oxford University Press.

Mahon, J. E. (2016). The definition of lying and deception. In E. N. Zalta (Ed.), *The stanford encyclopedia of philosophy*. Retrieved June 19, 2023, from https://plato.stanford.edu/archives/win2016/entries/lying-definition

McMyler, B. (2011). *Testimony, trust, and authority*. Oxford University Press.

Moran, R. (2005). Getting told and being believed. *Philosophers' Imprint, 5*(5), 1–29.

Rini, R. (2017). Fake news and partisan epistemology. *Kennedy Institute of Ethics Journal, 27*(2), 43–64.

Rupar, Aaron. 2020. Trump says 'sarcasm' was behind his coronavirus comments on disinfectant. Vox, 27 April 2020. Retrieved June 18, 2023, from https://www.vox.com/2020/4/27/21238141/trump-sarcasm-noble-prizes-hamberger-twitter-meltdown.

von Thun, S., & Friedemann. (1981). Miteinander reden: Störungen und Klärungen. In *Psychologie der zwischenmenschlichen Kommunikation*. Rowohlt.

The Daily Show. (2020). Jordan Klepper vs. Trump supporters: The complete collection. In *The daily social distancing show*. Collection from the daily show, 59:22. Retrieved June 16, 2023, from https://www.youtube.com/watch?v=7OeeHz0uNdM

Williamson, T. (2002). *Knowledge and its limits*. Oxford University Press on Demand.

Zagzebski, L. T. (2012). *Epistemic authority: A theory of trust, authority, and autonomy in belief*. Oxford University Press.

Describing the Post-truth Predicament

Politicization, Signaling, and the Epistemic Landscape

Casey Doyle

Contents

Abstract Political epistemologists, social psychologists, and others have sought to make sense of the prevalence of beliefs that conflict with expert consensus on politically charged topics. This chapter begins by making the case that this project would benefit from understanding how the process of politicization affects the epistemic landscape. It provides a model on which a politicized landscape is one in which one's stance on an inherently nonpolitical topic signals a political stance and, given one's assessment of the costs, it makes sense to engage in socially adaptive rather than truth-tracking cognition on the nonpolitical issue. The paper ends by considering how this bears on trust and the influence of social media on political opinions.

C. Doyle (✉)
Binghamton University, Binghamton, NY, USA

M. M. Resch et al. (eds.), *Trust and Disinformation*,
https://doi.org/10.1007/978-3-031-96790-0_5

1 A Puzzle about "Political" Opinions

A common trope in contemporary applied epistemology and political philosophy is to begin a piece with a list of examples of outrageous things people believe, for example: that Hilary Clinton sold weapons to ISIS (Snopes, 2016), that COVID is a government controlled bioweapon; that it is spread by 5 g towers; that pharmaceutical companies are intentionally spreading the disease (Celia et al., 2022), or that climate change is a hoax perpetuated by the Chinese government.[1] These examples are taken to be representative of a *crisis* in contemporary society, a crisis of trust (Rini, 2017, Hannon & de Ridder, 2022, Grundmann 2021; Levy 2022). Conspiracy theories are the most arresting, but not the only cases. Many resist the consensus amongst experts about climate change, vaccines, and masks. They hold what Levy (2022) calls "bad beliefs", a belief that "conflicts with the beliefs held by the *relevant epistemic authorities* and… despite the widespread public availability either of the evidence that supports more accurate beliefs or of the knowledge that the relevant authorities believe as they do" (ix). Many claim that social media and other technologies play a significant role in all of this (Rini, 2017, Joshi, 2021; Dutilh Novaes & de Ridder, 2019).

But people believe all sorts of zany things, and presumably always have. I believe, to some degree, that one can "jinx" good fortune by speaking about it. Countless teenagers are led to believe they can be whatever they want in life. And however outrageous the average COVID conspiracy theory is, it fails to stretch one's imagination in the way that garden variety religious beliefs do. Two features of the relevant cases are thought to call out for special explanation. First, people seem to be irrational in failing to trust relevant experts. Second, the discussions have focused on people's *political opinions*.[2]

But what are political opinions? Are the paradigmatic cases political? It is hard to see how the belief that masks are ineffective against the spread of COVID or that the vaccine is a means of implanting microchips are themselves political in any but the broadest sense. While conspiracy theories are typically political because they tend to implicate the behavior of political actors and public institutions (Keeley, 1999), often their original target claims are not. Many "bad beliefs" are about political actors and institutions, but many are not. Indeed, the more explicitly political the content of a belief — for example, beliefs about how to vote or what policies to enact — the less obvious it is that citizens of a democracy ought to defer (van Wietmarschen, 2018).

Although it is easy enough to feel discomfort about paradigmatic cases — to feel, that is, that something has gone wrong with people's beliefs about vaccines, masks, climate change, and so on — it is far from obvious just what the phenomenon is, unless it is simply that people can come to believe lots of silly things through diverse means and that this can have harmful consequences.

[1] Rini, 2017 and Levy, 2022 are good examples.

[2] See for example the essays in Bernecker et al., 2021 and Hannon and de Ridder 2022.

There is a puzzle here. On the one hand, it seems that the political character of these cases is what unifies them. On the other hand, it does not seem that there is anything inherently political about the content of the beliefs themselves. A natural way to resolve this is to recognize that these are cases in which an issue has *become politicized.* In the first instance, the question of masks concerns their efficacy and my decision as an individual to wear one. But this question has taken on a political dimension, which leads to trouble.[3]

It is plausible, then, that an understanding of how politicization affects the epistemic landscape will help us to understand some of the cases that trouble us. Advancing this understanding is the goal of the present chapter. On my view, a politicized environment is one in which taking a stance on some topic signals a stance on some political issue — typically support for a political party or side — and it is rational, given one's assessment of the costs, to consider the practical costs and benefits of signaling in this way. In a politicized environment, deliberation becomes a matter of social signaling rather than truth-tracking cognition.

In the next section I draw the distinction between truth-tracking and socially-adaptive cognition. Section three offers a characterization of politicization based on the idea of signaling. Section four surveys evidence indicating that, for most of us, political opinions are a matter of social signaling rather than narrowly truth-tracking cognition. After responding to objections, I conclude by considering the consequences for trust and the spread of disinformation through technology.

2 Truth-Tracking and Socially-Adaptive Cognition

Suppose that I am sitting on my chaise lounge reading *What we Owe to Each Other* when my children walk in and say the following: "Dad, we want to play ice hockey this year. Is it safe enough for us to play?" I put to myself the question "is ice hockey safe?" and consider the evidence that is available to me. Reflecting on this, and considering the relevant context (for example, comparing hockey with other sports they might play), I conclude, yes, it is safe enough.

This is an example of what I am going to call "truth-tracking cognition". I am faced with a question, answer it by taking a stance, where that stance is determined only, at least consciously, by my reflection on the evidence bearing on the truth of the target proposition. The idea of truth-tracking cognition is familiar. It is the kind of thing Jerry Fodor is talking about when he says that "the proper function of cognition [...] is the fixation of true beliefs" (2001, 68). Truth-tracking cognition is a form of thinking that aims at belief, where belief is understood as a propositional attitude that is constitutively governed by the truth norm. Belief "aims at the truth" (Williams, 1973, Velleman, 2000; Shah, 2003); to recognize that a belief is false is

[3]The present discussion can be seen as addressing an instance of a feature of contemporary cultural life that has raised concerns, namely that "everything is political", that politics infects too wide a range of activities in public (Talisse, 2019).

to immediately find fault with it and, if it is one's own, to discard it. Because beliefs are governed by this norm and are responsive only to evidence, one cannot "believe at will" (Williams, 1973). It is *irrational* to believe without sufficient evidence because evidence indicates the truth of a belief. All of these claims are controversial and in need of specification, of course, but the general picture is all we require here.

Now consider another case. The kids come in, but this time they say: "Dad, we want to play ice hockey this year. Is it safe enough for us to play? Mom says that it isn't." Things are quite different here. A whole set of additional considerations weigh on me. It is important that my partner and I represent a unified front against the cascading demands of the children. It is important that they maintain confidence in our assessment of risk. It is important that we all get along. And so on. Given that my partner's view on the matter is the cautious one, and the costs of getting things wrong if I agree aren't serious, I allow these concerns to weigh on me and decide to go along. I say "I agree with Mom. It isn't very safe."

This is what, following Williams (2019), I will call "socially adaptive cognition". In addition to serving a role akin to private maps that guide an individual through the world (Ramsey, 1990), the stances we take on the world are objects of social scrutiny. They matter to other people, and we consider the ways in which they matter when determining what stances to take. As Eric Funkhouser (2017) has put it, our stances can serve as "signals" to others: they carry information about us that is useful in social circumstances. We are aware of this and can use the stances we take to send the signals we want. We do this when we lie, but not only then. I am not lying when I tell my children that hockey is not safe. I have no settled opinion about it one way or the other.

Some hold that there is a single state, belief, that serves two functions: tracking the truth and social signaling (Funkhouser, 2017, Williams, 2019; Hannon & de Ridder, 2022). I deny this. As mentioned, belief is constitutively governed by the truth norm. Consequently, it is the product of conscious cognitive processes responsive exclusively to considerations bearing on the truth. If you are considering what you signal about yourself by taking a particular stance on the issue, then you just aren't reasoning toward belief. Instead, socially adaptive cognition yields a different attitude or set of attitudes. I prefer to use Cohen's (1992) term "acceptance" for this class. Acceptance involves undertaking a policy to affirm that p for practical reasons. This is what I'm doing when I take the stance that hockey is not safe. Thus, I use the term "stance" for the genus of which there are at least two species, belief and acceptance.[4]

This means that reasoning toward belief is exclusive. To the extent that you are engaged in it, socially adaptive considerations ought not get a grip on your conscious thinking. However, the evidence can weigh on one, along with considerations about social signaling, when reasoning toward acceptance.[5] Socially adaptive cognition is permissive. When reasoning toward acceptance, practical

[4] Precedents for this include Dennett's (1978) distinction between belief and opinion and Mercier's (2020) distinction between reflective and intuitive judgment.

[5] Social signaling is not the only kind of consideration that can motivate acceptance, of course.

considerations can play a direct role in lower-order reasoning. But at the higher-order, practical reasoning can make a difference to *which* state one aims at— belief or acceptance— and so which activity one performs. One faces a *decision,* even if it isn't entertained consciously, whether to reason toward belief or mere acceptance. In deciding this, one decides whether to *bracket* practical considerations, including social signals. Presumably, if the cost of getting things wrong is high then the rational thing to do is reason toward belief, bracketing the practical considerations (Papineau, 2013). If my partner had told the children hockey *is safe*, then things would be different. The costs of being wrong were I to agree with her might be high, so I should probably restrict myself to the evidence. However, if the costs of getting things wrong are low, it might be reasonable not to bracket practical considerations.

Three more points. First, I don't suppose that we are infallible about which form of cognition we are engaged in. You might take yourself to be engaged in truth-tracking cognition, restricting yourself to the evidence, when you have been moved by practical considerations. We can get these things wrong. This is especially plausible given that socially adaptive cognition is permissive. Second, the norms that govern truth-tracking inquiry do not apply in the same way to socially adaptive cognition. If I am pretending I am a dinosaur, it is no strike against my thinking that norms about deferring to the leading paleontologists about relevant aspects of dinosaur behavior do not get a grip. Third, as the above should make clear, whether it is rational to engage in socially-adaptive or truth-tracking cognition depends on the agent and their context. When my children added "Mom says it isn't" the context shifted in a way that explains my engaging in socially adaptive cognition.

This allows us to formulate the main claim to be defended here: when a topic becomes politicized the environment changes in such a way that it becomes intelligible for some (perhaps many, perhaps most) to go in for socially adaptive cognition when answering a question that, previously, would have been the subject of truth-tracking cognition.

3 Politicizing as Instituting Signals

As I will understand it, politicization occurs when formerly non-political actions come to possess a political meaning, so that their performance signals allegiance to a political side or identity. Importantly for our purposes, such actions can include taking stances, through assertions or other speech acts. Thus, affirming a proposition that might seem non-political can come to be an action with political meaning and signal a political identity.

Uses of "politicization" are broader than this, so it will help to locate our usage within that terrain. To politicize an issue is to make it come to have political significance. On one usage this is viewed as a good thing. For example, attempts were made to "politicize rape" by making people aware of the injustices of the existing laws and bring about change. In line with this the OED tells us that to politicize is

"To make political, esp. to make (a person, group, etc.) politically aware or politically active". The sense I have in mind tracks usage in contemporary political discussions in the US, and presumably elsewhere, where it is viewed pejoratively. To politicize an issue is to muddy the waters, to hinder clear debate and discussion, and to use a current crisis for gain. Often accusations of politicization are insincere, as when Republicans in the US accuse their opponents of "politicizing gun violence" (Blum, 2022). But it needn't be, as with politicizing climate change or masks. Obviously, politicization as such is neither good nor bad. Some subjects that are considered "private" should be appreciated in their full political dimensions, and vice versa. And, of course, reasonable people can disagree about this. The characterization of the effects of politicization on the epistemic landscape on offer here is neutral, though I aim to explain why it is a bad thing when it is.

An example will help. In a 2014 study, Kahan et al. found that people's risk perceptions correlated with their political orientation in a wide range of cases. Conservative Americans viewed marijuana legalization as dangerous while global warming and gun ownership were viewed as low risk, with liberals holding the opposing views. However, at the time, childhood vaccination was not strongly correlated with either political identity. This changed with COVID-19. Recent studies have shown that vaccine hesitancy has become correlated with political identity (Kirzinger et al., 2021). It has become politicized.

Suppose that in 2013 you learn from a coworker that they do not vaccinate their children. What does this tell you about them? Presumably, a fair amount. But if Kahane et al. are right, it doesn't indicate whether they are conservative or liberal. Fast-forward ten years, and things are different. When your coworker tells you that they do not vaccinate, this carries an additional message, namely, that they are conservative. That is, the utterance, the literal meaning of which only concerns vaccines and certain people, also carries a message about the speaker's political identity. This is not a matter of conversational implicature. But like conversational implicature, it can be cancelled. (e.g. "I'm not a conservative, but I worry a bit about this vaccine.")

There is nothing distinctive about "stances" here. The same broad phenomenon occurs with consumer choices, aesthetic preferences, where one lives, and more besides. An act that has no inherent political significance comes to serve as a political communicative act thanks to what Robert Talisse calls "saturation". Talisse describes the by now familiar phenomenon of physical and virtual sorting: the way choice over lifestyle and media leads people to cluster along political lines.

> The resulting clusters of individuals in homogeneous physical and social spaces reflect their political divisions. As a consequence, demographically and geographically sorted social spaces are also sites of political like-mindedness. (Talisse, 2019, 82).

In such a context, a non-political act—buying coffee, driving a car, even recycling—reliably indicates one's political identity. Those in the know can then interpret the act as an expression of the relevant political outlook, regardless of the intentions of the agent in question. According to Talisse, as a result:

> Sorting has wrought the convergence of politics with our broader identities, and so our partisan identities spill over into the facets of day-to-day life that otherwise would be regarded as non-political. (ibid., 89).

Drinking a particular cup of coffee is not inherently political. But if that coffee is a reliable indicator of a political identity, it signals this to people in one's environment.

If carrying a rifle in one's car becomes politicized, then the act of doing so communicates to some relevant audience that one endorses a particular political outlook. Likewise with wearing masks, getting vaccinated, or, in principle, anything else you like. And what goes for these overt physical actions applies to stance taking, as well.[6] Politicization of a topic is the process whereby a stance on some non-political issue comes to be saturated with political significance in just the way that other non-political acts can come to function as political communication. In a politicized environment taking a stance on what might seem a nonpolitical issue signals a stance on some political issue. And just as the political meaning of a cup of coffee might give you reason not to buy it; the political meaning of a stance can give people reason not to take it.

While we have focused on the way in which politicization interferes with deliberation about nonpolitical matters, it can also interfere with scientific inquiry. Maya Goldenberg, following Pielke, 2004, describes how widespread adoption of a "linear" model of the relation between science and politics—on which the results of evidence-based, allegedly value-neutral science directly entail verdicts about policy— leads people to treat debate about science as a proxy for policy debate. Focusing on climate change, she writes:

> Because of the scientized nature of the political debate, scientific studies showing meaningful connections between greenhouse gas emissions and actual projected climate changes are quickly interpreted to be supportive of actions to reduce emissions. Similarly, studies that cast doubt on the significance of such connections are interpreted as challenging the need for action. (Goldenberg, 2021, 101).

Here a failure to divorce political and scientific issues leads people to equate a stance on the science as itself a stance on policy. This impoverishes scientific debate and also leads to an impoverished sense of the political options.

4 Political Opinions Are Not Beliefs

So far we have a model for how politicization can affect the epistemic landscape. The question whether p arises for you. In the ordinary run of things, the thing to do is engage in truth-tracking cognition, restricting one's deliberation to considerations bearing on the truth of p. (Should I wear a mask in public? That depends on the risks, my comorbidities, what the experts are saying, how reliable they are, and so

[6] See Funkhouser (2017) for defense of the claim that propositional attitudes themselves function as signals.

on.) In a saturated political environment, affirming that p signals a particular political identity. This could be allegiance to a political party, figure, or broad ideology. To affirm that p, in one's context, is to mark oneself as a particular kind of person. In context, one has reasons for or against marking oneself in this way. This process could occur unintentionally, as a result of sorting or a bad model about the relation between science and politics, or it could be intentional, the product of demagogues and talking heads. Politicization is an impediment to clear thinking because it makes non-epistemic considerations germane that previously were not.

Surely a big part of escaping adolescence is learning to resist the spell of what one's actions and stances "mean" in this way. We all, to varying degrees, manage this. Why not think people do manage it when it comes to politically charged matters? In this section I'll survey evidence suggesting that people's opinions on these matters are not beliefs. That makes it plausible that their stances are responding to broadly practical considerations, including social signals, plausible, that is, that the model we have arrived at applies to the cases that concern us.

In "The Thought", Frege remarks that any declarative sentence type that can be uttered with assertoric force can be uttered without it.

> This happens when we do not speak seriously. As stage thunder is only apparent thunder and a stage fight is only an apparent fight, so stage assertion is only apparent assertion…. Therefore it must still always be asked, about what is presented in the form of an indicative sentence, whether it really contains an assertion. And this question must be answered in the negative if the requisite seriousness is lacking. (Frege, 1956, 294).

On the assumption that assertion is the expression of belief (Williams 1999), the absence of requisite seriousness would indicate the absence of belief. When it comes to many people's political opinions, it is hard to escape the suspicion that the requisite seriousness is often lacking, that they don't, or couldn't, believe what they say. Two features of belief stand out. First, they are robust states that typically persist over time. Second, they motivate action when coupled with relevant desires.[7] But our political opinions— by which I mean to include beliefs about politicized topics— aren't like that.

Notoriously, 15% of Trump supporters claimed of pictures of Trump and Obama's inaugurations that Trump's showed more people (Schaffner & Luks, 2018). Half of Republicans claimed there was something true about the "Pizzagate" conspiracy theory (Blake, 2016). One in seven Americans affirmed that "Obama is the antichrist" (Harris, 2013) while one in three affirmed that he was not born in the United States (Uscinski & Parent, 2014). While it seems that, when polled, people believe extraordinary things, they routinely give less partisan and more accurate answers when offered a small monetary incentive (Bullock et al., 2015). This has been taken as evidence that answers to such questions don't really represent beliefs but rather function as a kind of cheerleading called "expressive responding". The question whether Obama is the antichrist is interpreted as a question of whether, and

---

[7] These are not intended to be controversial claims about the nature of belief. They distinguish belief from nearby propositional attitudes, like pretense and presupposition.

how strongly, one disapproves of him. Furthermore, a substantial proportion of survey responses appear to be constructed on the spot (Zaller & Feldman, 1992).

Because they arise in the context of polling, it might be thought that these examples don't generalize. But like expressive responding and constructed opinions ordinary political opinions are "shallow" in the sense that they are easily changed and do not influence behavior.

Commitment to a political stance seems to shift with the position of the party, figure, or cultural group one aligns oneself with. A prominent example includes Republicans in the US shifting in their views on Isolationism with the rise to prominence of Donald Trump within the party (Pew, 2017). More generally, choice blindness studies have shown that people struggle to keep track of their political opinions over time. In Sweden, people were asked about their attitudes on policies promoted by different political parties, ranking degree of agreement on a 100 mm line (Hall et al., 2013). After some time, their rankings were presented back to them and they were asked to justify their choices. In some cases, the original ranking was presented. Others had their "opinion" manipulated, in that the ranking they were shown deviated significantly from that which they had originally offered. (Very roughly: You originally rank yourself as in strong agreement with a liberal party. Time passes, and you are shown a ranking of disagreement with the liberal party and told that this was your choice.) Only 22 percent noticed that their ranking had shifted and more than 90 percent offered a justification for the manipulated opinion.

Nor do politicized opinions motivate action in the way one would expect if they were beliefs in the ordinary sense. Hugo Mercier (2020) provides the example of a rumor spread in the town of Orleans in 1969 that Jewish shopkeepers kidnapped local girls and sold them into slavery. People appeared to accept this story, and yet they did not act in the way one would expect people who believed this to act (e.g. by charging into the store or alerting the authorities). Indeed, cases where people *do* act on their professed extreme political opinions stand out as noteworthy. Many Republicans claim that the 2020 US Presidential election was stolen and that their democracy was under attack, but only some stormed the Capitol building in January 2021. Given that people so rarely act on their political opinions, it is plausible to suppose that their primary function is social signaling.

In addition to anecdotes and empirical evidence, there are good a priori reasons to think that political opinions would be undertaken without the requisite seriousness. Bloom puts the point as follows:

> If I have the wrong theory of how to make scrambled eggs, they will come out too dry; if I have the wrong everyday morality, I will hurt those I love. But suppose I think that the leader of the opposing party has sex with pigs, or has thoroughly botched the arms deal with Iran. Unless I'm a member of a tiny powerful community, my beliefs have no effect on the world. This is certainly true as well for my views about the flat tax, global warming, and evolution. They don't have to be grounded in truth, because the truth value doesn't have any effect on my life. (Bloom, 2016, 236).

In many politicized domains, the costs of getting it wrong are not very high, which makes it intelligible not to bracket practical considerations and be moved by the signal sent by a particular stance. Likewise, many political topics are very complex and the costs of inquiring with the requisite seriousness are quite high while the rewards are low. But sending the right signal to my community is low cost and high reward (Huemer, 2016). In such a context it is instrumentally rational to refrain from engaging in truth-tracking inquiry and opt for mere signaling. This fits the model for how people make decisions about political matters offered by Achen and Bartels:

> …voters choose political parties, first and foremost, in order to align themselves with the appropriate coalition of social groups. Most citizens support a party not because they have carefully calculated that its policy positions are closest to their own, but rather because 'their kind' of person belongs to that party. (Achen & Bartels, 2016, 307).

Following Van Leeuwen's recent work on religious belief (2018), we can sum this up by saying that, for most of us, political opinions primarily play an "identity-constituting" role, rather than a truth-tracking and action-guiding one. Likewise, our reasoning about politicized matters is best understood as what Kahan calls "identity-protective cognition" (cf. Kahan et al., 2007). We take a particular political stance because doing so is part of the repertoire of being a particular kind of person.[8] Doing so signals to people that we are that kind of person. If this is true of political opinions generally, then it is plausible to hold that it is true of our stances on politicized topics.

5 Objections

Before considering what this story about politicization and the epistemic landscape tells us about trust and social media, I will respond to three objections.

5.1 *Shallow Stances Are Beliefs*

Levy (2022) argues that political opinions are beliefs that are formed rationally, by deferring to those that social cues indicate are trustworthy. On the view developed here political opinions are to a large degree stances responsive to practical considerations, primarily the signal sent by adopting them. On Levy's view, they are beliefs based on evidence. So who is right?

Levy accepts much of the evidence surveyed above, he also accepts that political opinions are shallow and responsive to what is approved by people like oneself. Yet he insists that they are beliefs because they are responsive to evidence and rationally

[8] Another relevant example is "clustering" (Joshi, 2021). Why is it that people who agree about foreign policy so reliably agree about abortion, the environment and tax policy? Would they so reliably converge in their stances on such disparate topics if they were primarily responding to the evidence?

formed. The disagreement turns on whether the property of being socially approved is, as Carter & McKenna (2020) put it, "epistemically irrelevant". Levy, following Rini (2017), argues as follows. From my point of view, those who are like me have a better track record on political and other issues. Thus, it is reasonable for me to conclude that they are more reliable than my political opponents. It is rational to favor their word and so to opt for the view that is favored in my milieu. Being socially approved is epistemically relevant. However, it has to be granted that being socially approved is only an indicator of the truth under certain narrow circumstances. Everyone should recognize that it would very often be a coincidence that the stances people around these parts approve of are the true ones (Greco, 2021). But it is dubious that our sensitivity to social signaling discriminates between the coincidences and the noncoincidences. (None of the evidence indicates that people are reflective about this at all, for example.) It is more plausible that, on political matters, we go in for what is socially acceptable regardless of whether there is independent reason to believe that what is acceptable is true. Thus, it is more plausible to suppose that political stance-taking is socially adaptive and not truth-tracking cognition.

5.2 More than Instrumental Rationality

I argued, following Bloom, Huemer, and others, that it is instrumentally rational for us to prioritize signaling when forming political stances given the costs of inquiry, the high cost of signaling poorly, and the low benefit of getting it right. One might worry that this takes on board too narrow a conception of rationality.[9] After all, inquiring into political matters and being informed about current events might be a matter of personal pride or a source of meaning in one's life. In line with this, one might agree with van Leeuwen that political opinions are identity constituting, but insist that it is political reflection and stance-taking that matters, rather than social signaling and approval.

It would surely be a mistake to suppose that people's interest in political reflection is justifiable only by the broad effects their actions have on their society. Trying to understand the issues of the day is, after all, simply an instance of trying to understand oneself and the world one lives in. The claim is not that the costs of getting things right, signaling, and the expected outcomes make it rationally *obligatory* to refrain from truth-tracking cognition. The point is instead the more modest one that, in such an environment, it is reasonable to expect that many people will opt for social signaling and few will feel the harmful effects of doing so. Furthermore, it should be noted that finding meaning in one's political outlook is consistent with social signaling, since community with people like oneself is a source of meaning. And finally, given that, as Jason Brennan puts it, "most people know nothing, and

[9] I thank David Coady for raising a version of this worry.

many people know *less* than nothing" (2016, 24), there is reason to doubt that the source of meaning is getting at the truth wherever it may lie as opposed to getting on with one's peers.[10]

5.3 The Costs of Getting It Wrong Are High

This raises what I take to be the central difficulty for the foregoing: getting it wrong about vaccines, masks, or climate change is not low cost.[11] So the rational thing to do is bracket practical considerations, including social signals.

Here are three responses. First, the paradigmatic cases we are interested in for the most part concern collective action problems. While getting it wrong is high cost for us, it is not high cost for me, considered in isolation. That suggests that we should think of socially-adaptive cognition in contexts like these as akin to freeloading. Second, people may have more confidence in their assessment of the social stakes of signaling than the costs of getting things wrong. For example, my confidence that my academic friends will shun me if I reject the consensus view on masks is much higher than my confidence about the efficacy of masks. Third, our assessments of risk appear to be the product of cultural cognition even for topics that are not politicized (Kahan et al., 2007). This might be rational, if one is deferring to one's peers, and has reason to believe that, with respect to what is dangerous and how to live, they have a better track record. Or it might not. It is widely recognized that people struggle with statistical reasoning. Rational or not, freeloading or not, it is to be expected that people will judge that high stakes public concerns are low risk for them, which renders socially-adaptive cognition intelligible from their point of view.

6 Trust and Technology

The view defended here has straightforward upshots for thinking about trust and the role of social media and related technologies on it.

Truth-tracking cognition aims at belief, is responsive to evidence, and is constrained by epistemic norms. One set of such norms concerns how to respond to the testimony of others, including experts. While there is debate about how to understand epistemic authority, it is widely accepted that it can be irrational not to defer to experts in some circumstances. Puzzlement about bad beliefs often begins by noting that many people violate these norms. But if reflection on politicized topics is socially-adaptive cognition, then these norms simply do not apply. From the point

[10] On political ignorance, see Somin, 2013.

[11] I thank Andreas Kaminski for raising this worry.

of view of the average political reasoner the word of an expert on climate change is akin to the word of a palaeontologist for a parent playing with their child: relevant but not authoritative. People are playing by the rules, they are just playing the wrong game.

For many people social media is the primary source of information about current events. This is worrying for a number of reasons, including the incentives of tech companies, the ease with which disinformation is spread, and the absence of fact-checking and moderation online. But the view defended here helps to bring out another, in some sense more obvious problem. In a slogan: social media is for social signaling. Posting, sharing, and in other ways engaging with "content" serve as signals to one's peers online. This renders social media an environment in which we should expect people to be hyper-sensitive to the social costs of the stances they take. And that makes it a poor venue for truth-tracking cognition. But things are worse than this. Social media algorithms weight different pieces of content based on expected engagement, drawing on data that function as a representation of what sort of person the user is. This encourages echo chambers, as has been noted. But it also serves to strengthen the signal between stances, on political and nonpolitical matters alike, and identity. In this way social media is a tool for politicization.

It should come as no surprise that social media interferes with treating difficult topics with the "requisite seriousness." The foregoing has provided a model for understanding this. Instead of individual irrationality or rational deference in a bad environment, many bad beliefs are, like political opinions generally, acts of social signaling. This raises different questions: is it rational for people to engage in social signaling in these contexts? How, concretely, does politicization happen and how can it be avoided for topics that are of broad social concern? How do we sever the connections between the narrowly political and nonpolitical topics to encourage open and evidence-based debate? These are questions for further research. If the argument of this chapter is along the right lines, it has indicated their relevance to discussions about the contemporary crisis of trust.[12]

References

Achen, C., & Bartels, L. (2016). *Democracy for realists*. Princeton University Press.

Bernecker, S., Flowerree, A., & Grundmann, T. (Eds.). (2021). *The epistemology of fake news*. Oxford University Press.

Blake, A. (2016). *Pizzagate' theory believed by nearly half of republicans: poll*. Washington Times. December 28.

Bloom, P. (2016). *Against empathy: the case for rational compassion*. Ecco.

[12] I presented an early version of this paper at the Trust and Disinformation Conference at the University of Stuttgart in 2022. Thanks to the audience and especially to Brian Ball, David Coady, and Andreas Kaminski for feedback after the talk. Thanks also to the students in my graduate seminar on trust at Binghamton University in spring 2023.

Blum, P. (2022). *Demands to not politicize tragedy are themselves political demands*. Washington Post. Retrieved September 4, 2022, from https://www.washingtonpost.com/politics/2022/05/26/demands-not-politicize-tragedy-are-themselves-political-demands/.

Brennan, J. (2016). *Against democracy*. Princeton University Press.

Bullock, J., Gerber, A., Hill, S., & Huber, G. (2015). Partisan bias in factual beliefs about politics. *Quarterly Journal of Political Science, 10*(4), 519–578.

Carter, J. A., & McKenna, R. (2020). On the skeptical import of motivated reasoning. *Canadian Journal of Philosophy, 50*, 702–718.

Celia, G., Lausi, G., Girelli, L., Cavicchiolo, E., Limone, P., Giannini, A. M., & Cozzolino, M. (2022). COVID-19 related conspiracy beliefs and their relationship with defense strategies, emotions, powerlessness, attitudes, and time perspective. *Frontiers in Psychology, 13*, 939615. https://doi.org/10.3389/fpsyg.2022.939615

Cohen, L. J. (1992). *An essay on belief and acceptance*. Clarendon Press.

Dennett, Daniel. 1978. How to change your mind. In his Brainstorms: philosophical essays on mind and psychology. : MIT Press.

Dutilh Novaes, C., & de Ridder, J. (2019). Is fake news old news? In S. Bernecker, A. Flowerree, & T. Grundmann (Eds.), *The epistemology of fake news* (pp. 156–179). Oxford University Press.

Fodor, J. (2001). *The mind doesn't work that way: the scope and limits of computational psychology*. MIT Press.

Frege, G. (1956). The thought: a logical inquiry. *Mind, 65*(259), 289–311.

Funkhouser, E. (2017). Beliefs as signals: a new function for belief. *Philosophical Psychology, 30*(6), 809–831.

Goldenberg, M. (2021). *Vaccine hesitancy: public trust, expertise, and the war on science*. University of Pittsburgh Press.

Greco, D. (2021). Climate change and cultural cognition. In M. Budolfson, T. McPherson, & D. Plunkett (Eds.), *Philosophy and climate change* (pp. 178–199). Oxford University Press.

Grundmann, T. (2021). Facing epistemic authorities: where democratic ideals and critical thinking mislead cognition. In S. Bernecker, A. Flowerree, & T. Grundmann (Eds.), *The epistemology of fake news* (pp. 134–155). Oxford University Press.

Hall L, Strandberg T, Pärnamets P, Lind A, Tärning B, Johansson P (2013) How the polls can be both spot on and dead wrong: Using choice blindness to shift political attitudes and voter intentions. *PLoS ONE 8*(4): e60554. https://doi.org/10.1371/journal.pone.0060554

Hannon, M., & de Ridder, J. (2022). The point of political belief. In M. Hannon & J. de Ridder (Eds.), *Routledge handbook of political epistemology* (pp. 156–165). Routledge.

Mercier, H. (2020). Not born yesterday: The science of who we trust and what we believe, Princeton, NJ: Princeton University Press.

Michael, H., & de Ridder, J. (Eds.). (2022). *Routledge handbook of political epistemology*. Routledge.

Harris, P. (2013). One in four Americans think Obama may be the antichrist, survey says. In *The guardian*.

Huemer, M. (2016). Why people are irrational about politics. In J. Anomaly, G. Brennan, M. Munger, & G. Sayre-McCord (Eds.), *Philosophy, politics, and economics: an anthology*. Oxford University Press.

Joshi, H. (2021). *Why it's OK to speak your mind*. Routledge.

Kahan, D., Braman, D., Gastil, J., Paul Slovic, C. K., & Mertz. (2007). Culture and identity-protective cognition: Explaining the white male effect in risk perception. *Journal of Legal Empirical Studies, 4*(3), 465–505. https://doi.org/10.1111/j.1740-1461.2007.00097.x

Keeley, B. (1999). Of conspiracy theories. *The Journal of Philosophy, 96*(3), 109–126.

Kirzinger, A., Kearney, A., Hamel, L., & Brodie, M. (2021). *KFF COVID-19 vaccine monitor: The increasing importance of partisanship in predicting COVID-19 vaccination status*. Kaiser Foundation. Retrieved September 2, 2022, from https://www.kff.org/coronavirus-covid-19/poll-finding/importance-of-partisanship-predicting-vaccination-status

Levy, N. (2022). *Bad beliefs: Why they happen to good people*. Oxford University Press.

Papineau, D. (2013). There are no norms of belief. In T. Chan (Ed.), *The aim of belief* (pp. 64–79). Oxford University Press.

Pew. (2017). Continued partisan divides in views of the impact of free trade agreements. Retrieved September 3, 2022, from https://www.pewresearch.org/short-reads/2017/04/25/support-for-free-trade-agreements-rebounds-modestly-but-wide-partisan-differences-remain/

Pielke, R. (2004). When scientists politicize science: Making sense of controversy over *the skeptical environmentalist*. *Environmental Science and Policy, 7*(5), 405–417.

Ramsey, F. (1990). In D. H. Mellor (Ed.), *Philosophical papers*. Cambridge University Press.

Rini, R. (2017). Fake news and partisan epistemology. *Kennedy Institute of Ethics Journal, 27*, 43–64.

Schaffner, B., & Luks, S. (2018). Misinformation or expressive responding? What an inauguration crowd can tell us about the source of political misinformation in surveys. *Public Opinion Quarterly, 82*(1), 135–147.

Shah, N. (2003). How truth governs belief. *The Philosophical Review, 112*(4), 447–482.

Snopes. (2016). Shots hired. Retrieved September 2, 2022, from http://www.snopes.com/wikileaks-cofirms-hillary-clinton-sold-weapons-to-isis

Somin, I. (2013). *Democracy and political ignorance: Why smaller government is better*. Stanford University Press.

Talisse, R. (2019). *Overdoing democracy*. Oxford University Press.

Uscinski, J., & Parent, J. (2014). *American conspiracy theories*. Oxford University Press.

Leeuwen, V., & Neil. (2018). The factual belief fallacy. *Contemporary Pragmatism, 15*(3), 319–343.

van Wietmarschen, H. (2018). Political testimony. *Politics, Philosophy & Economics, 18*(1), 23–45. https://doi.org/10.1177/1470594X18798062

Velleman, J. D. (2000). *The possibility of practical reason*. Oxford University Press.

Williams, Bernard. (1973). Deciding to believe. In his *Problems of the self*. Cambridge University Press, 136–151.

Williams, Bernard. (1999). *Truth and Truthfulness*. Princeton University Press.

Williams, David. (2019). Socially adaptive belief. *Mind & language 36* (3):333–354. Retrieved September 2, 2022, from https://doi.org/10.1111/mila.12294

Zaller, J., & Feldman, S. (1992). A simple theory of the survey response: Answering questions versus revealing preferences. *American Journal of Political Science, 36*(3), 579–616.

Fake News and Conspiracy Narratives in the Context of Regimes of "Posttruth": An Analysis from the Perspective of Nihilism

Oliver Zöllner

Contents

Abstract This chapter analyzes why fake news and conspiracy narratives have been able to embed themselves in much of digital everyday life. To this end, it will trace how concepts of truth have increasingly faltered in modernity, as well as the normative consequences of a resulting crisis of truth. The chapter's central thesis is that disinformation, lies, and fake news are expressions of a nihilistic attitude in society. Focusing particularly on conspiracy narratives, right-wing populist media messages, and constructs of the untrue, the article details the extent to which shocks to the presuppositions of knowledge, and a nihilism based on kitsch – the sticky falsehood that seems so attractive – are observable as a system of posttruth.

Portions of this article, in different form and in German, first appeared in the author's long-form essay "Klebrige Falschheit: Desinformation als nihilistischer Kitsch der Digitalität", published in *Digitalisierung und Demokratie: Ethische Perspektiven* (eds. Petra Grimm and Oliver Zöllner, 65–104. Stuttgart: Franz Steiner Verlag 2020), used here with the kind permission of the publisher.

O. Zöllner (✉)
Stuttgart Media University, Stuttgart, Germany

Heinrich Heine University Düsseldorf, Düsseldorf, Germany
e-mail: zoellner@hdm-stuttgart.de

M. M. Resch et al. (eds.), *Trust and Disinformation*,
https://doi.org/10.1007/978-3-031-96790-0_6

1 Introduction

The media systems in industrial societies, and with them the patterns of their usage, have changed considerably over the last 30 years. The rapid market penetration of digital media technologies from around 1995 onwards has led to an enormous multiplication of digital outlets as well as an emergence of new publics (see Fuchs, 2023, 221–43, Habermas, 2023 and Hjarvard, 2008). The Internet and social networking sites are by now ubiquitous and have led to new cultural practices of interaction that were hardly conceivable before digitalization (see Stalder, 2018). In the context of a progressing "deep mediatization" (Hepp, 2020), virtually anyone can now become a public communicator or "influencer" with relatively little effort, and employ a wide range of digital devices to produce messages and disseminate them to dispersed audiences worldwide.

Enabling this process, new kinds of authorities have established themselves in an astonishingly short time, among them large technology companies such as Alphabet, Apple, Amazon, Meta and others with their various platforms and their underlying algorithms. In a context of disruption by technology (see Taplin, 2017), they create new patterns and habits of media usage as well as new modes of producing and receiving information (see Gertz, 2024, 137–60 and Neuberger et al., 2019). As intermediaries, these corporations, to a considerable extent, shape the institutional framework for public communication in society. Indeed, they enable the spreading of a vast amount of content (some of it factual, some not) in multifarious ways as had been unthinkable a mere twenty years earlier.

One story from Germany may serve as an example of how the untrue finds its way to audiences worldwide via digital media. That minor, but typical everyday story was published by the right-wing populist US website Breitbart News on January 3, 2017. Under the headline "Revealed: 1,000-Man Mob Attack Police, Set Germany's Oldest Church Alight on New Year's Eve," author Virginia Hale writes that a mob of more than a thousand Muslim men set fire to "Germany's oldest church" in Dortmund on New Year's Eve, shouting "Allahu akhbar" (God is great) and attacking police officers (see Hale, 2017). The story quickly went viral worldwide, especially on social media, and even found coverage in some otherwise reputable news outlets.

Dortmund police were quick to clarify that there had been no such incident. An official statement described a rather quiet course of events: "No outstanding or spectacular New Year's Eve incidents have been reported to date." There had been, in sober police language, only "some inappropriate use of New Year's Eve fireworks" (see Polizei NRW Dortmund, 2017a, 2017b; trans.). In fact, a single out-of-control firecracker scorched a building site's safety net at Reinoldi Church, afflicting only minor damage to the property. There was no Muslim "mob" and the church in question is by no means the oldest house of worship in Germany. Breitbart News supplemented the original story a few days later with an abbreviated statement from police that it had been a quiet night, but otherwise stood vehemently by its story. This item can easily be identified as "fake news" (see Lazer et al., 2018) whereby the recipients are being "invited to join a shared fantasy" (O'Shaughnessy, 2020, 55). The article's core message is a fabrication, the blurry picture adorning it is out of context.

The story of the allegedly oldest church in Germany allegedly set on fire by Muslims thus claims to be "news" despite the factual refutation of all essential story elements. It may, however, aspire to be accepted as "news," at least by some interested target groups, precisely because of the factual refutation of its core story elements, since that factual refutation was made by an entity that Breitbart News (and similar organizations) deny the legitimacy to define the truth in the first place: that is, state organs, represented here by the local police authority. In the era of "post-truth," seemingly anything can be written and published based on idiosyncratic and self-sufficient ways of thinking, as the Breitbart News Story shows, thereby discursively creating "new realities" in order "to serve a political message" (Kalpokas, 2020, 71). The technical means for this endeavor are available to almost anyone in digital form and are easy to use. Accordingly, anyone can now create their own truths and disseminate their own worldviews.

The terminology for this distribution of falsehood is somewhat blurred. Disinformation, false information, fake news, rumor, and propaganda are not the same thing, but in the Internet age they lead back to a common question: what is "news," what is "truth," who defines it? The sheer increase in the amount of information available has obviously shaken the already problematic idea of truth (see Blackburn, 2005): "More information produces not more clarity, but more confusion" (Bridle, 2018, 188). In a kind of macro-perspective, the first question to ask against this background is: how did it come to this?

2 The Systemic Plumping Up of the Untrue

According to media theorist Vilém Flusser (1996), human beings try to find back to their anchoring in the world, and thus to their individual point of view, by means of "technoimages," or technical images (see Vuger, 2021). Such images can be both human- and computer-generated. The world of this technoimagination is, as it were, 'post-alphabetic' and thus geared towards the logic of the digital world. The Internet with its manifold text–image–sound combinations may be seen as an essential example of these new forms of argumentation and world indexing. The generated technoimages and their codes are, according to Flusser, "extraordinarily dangerous: they program us without having been seen through in their essence, and thus threaten us as opaque walls instead of connecting us to truth as visible bridges" (1996, 105; trans.). The technical images design "meanings on deceptive surfaces [...]" (Flusser, 1990, 45; trans.). Accordingly, disinformation in all its varieties so frequently disseminated via the digital sphere seems like an expression of a crisis of social communication. Since technoimagination is no longer a form of classically, linearly organized form of communication, and in a similar vein is no longer oriented towards objectivity, the search for truth in such a context no longer appears as a "voyage of discovery" but as an attempt to come to an intersubjective agreement with others about what constitutes the world (Flusser, 1996, 213). Society, in this perspective,

> no longer seeks truth in order to know and dominate the world, but to be able to live in it together with others. The consequences of such an upheaval are unimaginable, [...] above all because human life is immersed in a radically new mood: Truth-seeking ceases to be 'epistemological' and becomes again a 'religious quest,' but in a new and as yet unimagined sense. (ibid.)

The constructs of the untrue thus serve as inflated news dummies of a communal search for meaning which, however, has been deprived of the foundation of what once could be regarded as meaningful. Truth gets imagined, but it increasingly lacks a reference back to an objectivity that can be agreed upon by all. Constructivism seems to be a sign of the times (see Blackburn, 2005, 124–28).

Harry G. Frankfurt (2005) uses the catchy image of two bulls producing faeces in an argument as a metaphor for the current treatment of truth in contemporary societies characterized by high levels of mediatization, marketing orientation, and impression management (see also Couldry & Hepp, 2016). In order to distract and impress its opponent, one of the two animals emits excrement. According to Frankfurt, what we colloquially call "bullshit" is the central image of a loss of the social agreement to refer to truth. Instead, in (post-, meta-) modern Western societies, new forms of non-truth take a prominent position, which, however, cannot be equated with outright lies: "[...] the essence of bullshit is not that it is *false* but that it is *phony*" (Frankfurt, 2005, 47; original italics). "Bullshit" has divorced itself from the fundamental concern for truth (ibid., 30). Its main purpose and goal is to create an impression that may be false and untrue, but that focuses on the person presenting the bullshit and thus distracts from the matter at hand (ibid., 14, 18). In a teleological-utilitarian tradition of thought, then, the focus is on the outcomes: "The end determines the means" (Waldvogel, 2018, 29; trans.). References to the much-used concept of impression management (see Harlow, 2018; Nassehi, 2024, 205), located at the overlap of social psychology and applied persuasive communication, are obvious here. As Frankfurt notes critically:

> The realms of advertising and of public relations, and the nowadays closely related realm of politics, are replete with instances of bullshit so unmitigated that they can serve among the most undisputable and classic paradigms of the concept. (Frankfurt, 2005, 22)

The bullshitter, not being a liar, simply does not care about the truth: "The main point is to make possible a high level of candour and an experimental or adventuresome approach to the subjects under discussion" (Frankfurt, 2005, 36–37). In a political context, the most prominent representatives of this style of untruthfulness may be leaders of state or heads of government such as Vladimir Putin, Boris Johnson, Donald Trump and others of this ilk. Perhaps most clearly, President Trump, known as a master of "alternative facts" (see Hellinger, 2019, 79–80; Neuberger, 2018, 45), during his first presidential turn most notably translated the bullshit concept into political practice on October 23, 2018. Asked by a journalist what evidence he had for his squishy claim that among the refugees from Central America then currently on their way toward the U.S. border were also shady people "from the Middle East," the president merely replied evasively but vehemently: "There is no proof of anything, there is no proof of anything, but there very well could be" (quoted in Hutchinson et al. 2018).

There is no proof, but there very well could be: It is hard to find a more apt and concise example of the loss of the *will to tell the truth*, or a better illustration of a *posttruth* style of politics and general patterns of thinking, uttered by a political actor known for "his assaults on facts and propagation of falsities" (Sotirovic, 2020, 511). The Brexit campaign with all its fanciful claims and embellishments could also be mentioned in this context, though in all of its complexity it should rather be seen as an expression of an elaborate conspiracy theory (or rather a 'conspiracy narrative' as such a construct is no theory in the strict sense of the term), demonstrating the overspill of fictional structures into entirely real politics with equally real consequences (see Bergmann and Butter, 2020; O'Toole, 2018). The Covid-19 pandemic and the many fanciful examples of fake news and conspiracy narratives attached to it are prime examples of such falsity in an exceptional situation of crisis on a global level (see Amlinger & Nachtwey, 2022, 247–98; Butter & Knight, 2023; Nocun & Lamberty, 2020, 252–74).

What Beck (2016, xi) in this context has called a world "gone mad," is, in Frankfurt's (2006) analysis of the truth foundations of our present, a world increasingly built on bullshit and lies in which untruths are purposefully deployed to drive us mad: "The world we live in, insofar as our understanding of it is fashioned by the lie, is an imaginary world. [...] So [lies] are intended, in a very real way, to make us crazy" (Frankfurt, 2006, 77–78). The world thus becomes more unstable and susceptible to further constructs of non-truth and disinformation. Humans are losing their moorings.

It can therefore be argued with Flusser (1996) and Frankfurt (2005, 2006) that in the present day and age there is an observable gradual shift away from a recourse to truth that is in principle considered desirable by some (although in functional subsystems such as investigative journalism, as well as in science and research, truth continues to be regarded as an essential standard), and that the dominant orientation of contemporary culture is by and large deviating in the direction of untruthful *bullshitting*, which can be seen as a simulation *beyond truth* and prone to showmanship. In the end, such tendencies may establish falsehood built entirely on untruthfulness and lying. It is not difficult to interpret the effect-laden simulation, or bullshit, as a reference to the well-known model of hyperreality as introduced by Jean Baudrillard (1978). Conceptually, it is about the "substitution of the real by signs of the real, that is, a dissuasive operation [...]" (Baudrillard, 1978, 4; trans.). This notion, in which the world is increasingly perceived and constructed through its media representations, and in which fictionalizations creep in, until (in part, at any rate) the world emerges as something hyperreal beyond truth and fiction, fits congenially in with Frankfurt's reflections on facticity and the debates about increasing levels of mediatization and dubious constructions of reality introduced earlier.

In this view, conspiracy narratives and other forms of disinformation acquire their own kind of reality in the public realm of discourse. While they may be refutable and falsifiable on the rational level, the act of their public dismantling by professional authorities only raises their level of distribution in the media sphere, and the steadfastness of their adherents (see McKenzie-McHarg, 2020 and Sunstein & Vermeule, 2009). The denial of rational factuality is, after all, the very essence of conspiracy-narrative-based disinformation.

Jayson Harsin (2015) calls such a systemic dominance of the untrue in a context of social and political discourse heavily influenced by social media and other online platforms a "regime of posttruth". He regards this phenomenon as a characteristic of "post-political" societies, i.e., an expression of a disruptive transition from traditionally organized politics to new and increasingly more ambiguous forms of political action for which there is apparently no better term than the one with the prefix '*post-*' (see also Ball, 2017). Both *post-politics* and *posttruth*, understood in their combination as a "political apparatus," are based on "'participatory' social media politics" (Harsin, 2015, 331). Harsin puts the participatory aspect of social networking sites in quotation marks, since the forms of collaboration and participation that can be observed in these spaces of discourse are, for him, ultimately mere illusions of participation. The exchange of arguments, texts and symbols – the technoimages in Flusser's (1996) terminology – takes place under very restrictive terms of use stipulated by the platforms, on their digital private home turf, as it were. The continuous acts of mutual observing, sharing, 'Liking' and commenting by online actors serves a political–economic purpose. In the system of "surveillance capitalism" as postulated by Shoshana Zuboff (2019), the large data sets produced on social media generate profit for the platform operators as well as define norms for individual orientation and behavior (Zuboff, 2019, 199–232). In this respect, Harsin's description of the post-political posttruth society of control fits in with Zuboff's view of a post-democratic society in which the permanently fueled production, and subsequent evaluation, of personal data as building blocks of dominion that is at best economically legitimized, replaces previous concepts of democracy, human autonomy, and responsibility (see Harsin, 2015, 330–33 and Zuboff, 2019, 351–75).

Against this background, conspiracy narratives and other forms of disinformation appear as relatively arbitrary vehicles for communication which (apart from their tangible persuasive and political-propagandistic goals) simply serve to produce interaction data for the purpose of their economic exploitation. In their guise as fake news, they pretend to serve the purpose of organizing fact-based knowledge, reminiscent of journalistically researched and verified information, but in fact they are mere reflections of falsehood that negate truth by simulating an alternative reality. *Posttruth* thus appears as an expression of a transitional state of truth, which, as noted above, is conceptually in distress and increasingly mutating into falsity. Such *post-truthful* communication can also be called puffed-up falseness. It is a kind of falseness that pretends to be true by imitating established forms of truthful messages. In this way, it resembles the tackiness of *kitsch*.

3 The Falseness of Kitsch

It is the nature of *kitsch*, i.e., exemplars of overblown reality in the form of figurines, figurations, or figments that stick to one's imagination and may even be appealing, either naïvely or as a guilty pleasure, to be based on falsehood and to merely simulate the real. Conspiracy narratives and fake news puff themselves up into absolute, hermetic models of world explanation that do not tolerate falsification (as would a scientific hypothesis or theory). We can easily identify such disinformation, following Hans-Dieter Gelfert (2000), as "ideological kitsch": "The fading out of everything that does not fit into the ideologically cemented total explanation of the world corresponds exactly to the mendacity that kitsch exhibits in all its varieties" (Gelfert, 2000, 60; trans.). Kitsch thereby always points back to the past, recycling it in a desirable way. As an aesthetic phenomenon, it is a "revenge on modernity" (Liessmann, 2002, 73; trans.). Accordingly, dictatorships, for example, employ it for their self-legitimation and historicization (see Heller, 2008 with examples taken from twentieth-century dictatorships and their branding). Kitsch is based on a pose that, akin to bullshit, places the producer in the center and discredits claims to truth.

This also works in the context of democratic societies. The sweet smiles of a British royal couple, caught in a cleverly launched (and by now 'iconic') PR photo arranged in a staged picnic setting, with its symbolic claim to the principle of monarchic rule (see Zöllner, 2019), can be just as false and untrue as the Breitbart News story about the church in Dortmund allegedly set alight but which really was not. In a similar vein, the fully computer-generated image of Pope Francis in a white puffer jacket that went 'viral' worldwide is a new form of falsehood (see Tolentino, 2023) that may shatter people's trust in the veracity of images and news content in general. Such stories, kitschy technoimages that they are, pursue different goals, but have in common that they oppose the idea of truthfulness. They thus reflect the state of the posttruth society in which they are produced. The staged photo of Prince William and his wife presenting their newborn offspring refers to the archaic narrative of fertility (Zöllner, 2019, 109) and is based on a premodern idea of societal organization. Conspiracy narrative-based fake news, such as the xenophobic story of the allegedly marauding mob of Muslim men in Dortmund, harkens back to archaic primordial narratives of human survival that is fueled by the fear of an alien invader who poses a potentially lethal threat to one's own group. Combined with historical and repeatedly recycled narratives of threat, such constructs may grow into complex retro-narratives (see Gadinger, 2019, 138; Müller, 2019).

Narratives of this kind tend to stick: they are constantly repeated, can be located in the furthest corners of the media system, and become fixed in the memory of recipients and within social discourses – either despised or applauded, whether in the hundred thousandth chemtrail video on YouTube (as an implicit longing for an unspoiled environment of the pre-industrial era), in the form of the umpteenth reiteration of "White Replacement Theory" (fear of immigration and loss of status), at demonstrations and rallies of some non-traditional political groupings and movements, or in online comment forums now so ubiquitous (see Precht, 2019). This kitsch is backward-looking, penetrating and sticky, but paradoxically, for that very reason, enjoyably consumable:

> … the sticky structure of kitsch is also experienced – albeit affirmatively – in the enjoyment of kitsch itself. In stickiness, a state of human existence symbolizes itself that we could call the *pleasurable experience of being.* (Giesz, 1971, 41; original italics; trans.)

Kitsch's main function here is "to create moods" (Giesz, 1971, 44; trans.). What opposes established values "is not recognized by kitsch as something different and alien, but demonized as the quintessentially negative" (Gelfert, 2000, 78; trans.). The authorities that can set such values are demonized right along with it, as can be observed in the many highly polarizing and ideologically inflected messages posted in discussion forums or on social networking sites that are far removed from sound and sober argumentation. "Therefore, in kitsch, just about every value is accompanied by its demonized counter-image, which constitutes its characteristic black-and-white painting" (ibid.). This apparently succeeds quite well in the news mock-ups described earlier, in which the world is always portrayed in a particularly simple but also antagonizing way: threats abounding just about everywhere, or so it seems. We may therefore conclude by taking a closer look at ideological kitsch and its inherent tackiness from the philosophical perspective of nihilism, which places this kind of negation at the heart of its argumentations.

4 Nihilism and Disruption

Bullshit thrives in duplication. The evolution of media systems, most prominently the multiplication of channels and outlets, provides the foundation for this expansion. Progressive mediatization multiplies bullshit. Truth becomes variable or flexible. The posttruth-oriented, puffed-up falsehood demands attention and even appears to many as credible information. Messages of this type often serve political-propagandistic purposes, but above all they are elements of a comprehensive system of economic exploitation of data. Against this background, truth is beginning to seem like a thing of the past. Society is all too often content with a simulated substitute, the kitsch (see Flusser, 1985). This, for the time being, is the setting for dealing with conspiracy narratives and the fake news that are so often based on them. They point at the end of falsifiable meanings with a linear logic, and the decline of the legitimacy of truth. Behind the production, distribution and reception of disinformation could lie a nihilistic principle.

Donald A. Crosby (1988), in his systematization of nihilism, identifies five types of this philosophical approach to understanding the world, among which are political nihilism which denies and seeks to radically sweep away the existing political, social, and cultural order of life, and epistemological nihilism which holds that there is no truth or meaning beyond individual beliefs or outside of precisely definable collectivities (see Crosby, 1988, 9–36). Common to all varieties of nihilism is an attitude of negation and denial of essential aspects of human life or social organization. Nihilism "is about evading reality rather than confronting it, about believing in other worlds rather than accepting this one, and about trying to make ourselves feel powerful rather than admitting our own weaknesses" (Gertz, 2019, 73). Political nihilism seems to be a considerable motivation for many radical political producers of fake news and disinformation. Breitbart News and its long-time editor, intermittent presidential chief strategist and political advisor Steve Bannon would be an example of this (see Lewis, 2018).

In the context of our debate, a political-epistemological variety of nihilism seems to be at play. It is related to the worldview of constructivism already introduced above, according to which reality and truth result from individual agreements of the actors involved, i.e., a radical relativism of truthfulness (see Blackburn, 2005, 124–28). The range and the power of reason are strongly curtailed in this perspective, for reason can only unfold within particular systems of belief and discourse. The verifiability or falsifiability of arguments is thus not given, since virtually every actor is 'right' within his or her context of argumentation. However, it also follows from this position that "all such systems are themselves ultimately arbitrary in the sense of being beyond rational criticism or support" (Crosby, 1988, 18). This arbitrary nature of reasoning, a rejection of classical epistemology, has been prominently analyzed by Friedrich Nietzsche (1988 [1887]). He posits that will is everything and permeates all human striving. The world, he argues, is a chaos of conflicting forces and wills; life manifests itself in a "hatred of what is human," an "abhorrence [...] of reason itself" (Nietzsche, 1988, 165; trans.), with which is associated a turning away from the will to truth (ibid., 163). Man would rather "want nothingness than not want at all ..." (ibid., 165; trans.). Thus, in a Nietzschean interpretation, one could surmise that truth is purely strategic and therefore destructive – and may thus turn against life itself (see also Blackburn, 2005, 75–106; Heinrich, 2009, 89 and Raffnsøe, 2007, 135–39).

Conspiracy narratives, fake news, and other constructs of ideological and political kitsch may be interpreted, in the context of such a nihilistic view, as expressions of paradoxical willful performances of mendacious truth that call life into question; an emphasis on a world without consistent facts and meaning; an expression of a negative longing for the controllability of what is actually uncontrollable; in short, a means of domination (see Crosby, 1988, 19). If we understand the producers of fake news, conspiracy narratives, and related informational mock-ups, insofar as they are private citizens (and not state propagandists), as "disenchanted serious persons," they appear to us as people who now reject the real serious world they once embraced and who now embrace meaninglessness instead, "for in a world with no external authority the disenchanted serious person prefers the annihilation of nihilism to the anxiety of freedom" (Gertz, 2019, 100).

It may be the fear of the impositions of a cosmopolitanized modernity, all its constant deep transformations, and precisely its freedom (see Beck, 2016) that drive the disenchanted to actively produce falsehood, to work with constructs of falsity – in other words, to change the world according to their will. This striving for change is equally aggressive and regressive: by confronting the world's falsifiers with freedom – their own as well as other people's freedom – they are burdened with responsibility, and this responsibility elicits fear in them. Facets of human existence are negated by the protagonists of falsity through their reevaluation of values, thus returning themselves to a quasi-infantile state: a state before adult responsibility, actualized as one of irresponsibility and unfreedom (Gertz, 2019, 99–100). This regressive pattern, moreover, also applies to the aesthetic fake: "Childlike innocence is a central value of kitsch" (Gelfert, 2000, 68; trans.). For a society committed to the values of modernity and enlightenment, this active nihilistic reevaluation in the form of regression is self-evidently highly problematic. If society at large can no longer agree on what is true, this appears as a "symptom of a deeper crisis of democratic societies" (Butter, 2018, 233; trans.).

For the distributors of such disinformation, i.e., the social media platforms or the tech companies operating them, these messages are nothing but lubricants for their business model. Data production leads to increased interaction between users, which in turn leads to larger data sets. The latter can be linked, analyzed and used for targeted advertising and behavioral predictions, among other purposes, as part of a comprehensive system of surveillance capitalism (see Zuboff, 2019 and Nassehi, 2024, 193–94). By pursuing these business models, the big tech corporations become, politically speaking, the overseers and controllers of the online as well as increasingly the offline lives of large parts of the population. From the viewpoint of active nihilism, they become "masters of radical indifference" (Zuboff, 2019, 504) who take advantage of the collectivist mindset of many people – their tendency to "herd formation" (Nietzsche, 1988, 136–37; trans.). Zuboff contextualizes 'herding' within the metaphor of the "hive collective" as an essential part of the predictive imperative of surveillance capitalism, which we can identify as political nihilism:

> For the sake of its own commercial success, surveillance capitalism aims us toward the hive collective. This privatized instrumentarian social order is a new form of collectivism in which it is the market, not the state, which concentrates both knowledge and freedom within its domain. (Zuboff, 2019, 504)

This approach to dominion is the end of the reciprocity that has so far been the basis of the model of liberal democracy. The nation state as we have known it until now, and within whose structural framework the world still seems to be organized, is gradually taking a back seat to the power of tech corporations, their digital platforms and the rules they set without democratic control, which ultimately calls human autonomy into question. Users are shown journalistically researched articles via their news feeds seemingly on an equal footing with 'colorful' gossip and tabloid stories and deliberately manipulated disinformation from dubious sources, on the whim of an algorithm, as long as there is a large enough grouping of 'friends' or followers that have read, liked or commented on those items. This means that "anyone who can set topics by deliberately infecting the discourse with data undermines the power circle of the political system" (Nassehi, 2024, 95–96). This process has implications for citizens' worldviews, perhaps even their voting decisions (see Stark et al., 2018). The interactions of users, their datafied behavioral traces, serve and reinforce the exploitation contexts of surveillance capitalism described earlier.

Organized, accountable politics with codified rules is thus replaced by a kind of technological priestly caste that arbitrarily directs the production, supply, and utilization of data on its platforms: "a privileged priesthood of 'tuners' rules the connected hive, cultivating it as a source of continuous raw-material supply" (Zuboff, 2019, 505). Without any explicit reference to Nietzsche, Zuboff thus neatly traces this author's theoretical outline of nihilism, according to whom a caste of "ascetic priests" exploits the "herd formation" of human beings for the purposes of their "will to power" (Nietzsche, 1988, 136–37 and Nietzsche 1994, 118; trans.). In the context of our argumentation, the caste of priests are tech corporations like Alphabet, Apple, Meta Platforms, Twitter/X, etc. with their often quasi-religious promises of salvation and product presentation (see Gertz, 2024, 137–60 and Lanier, 2013, 124–32), as well as individual "tech evangelists" working as company leaders or publicists who preach perpetual progress and the next technological update. Fine examples of this tendency are the by now legendary charismatic appearances of the late Steve Jobs, CEO of Apple, Inc. (see Isaacson, 2011), entrepreneur Elon Musk's public musings, or Kelly's (2016) writings that suggest that there is no alternative to progress-through-technology. The inherent "will to power" of this kind can be found in contemporary industrial societies in a political–economic reading as the narrative of disruption, understood as a process aiming at ultimately destroying a given status quo from which an innovation, real or purported, eventually results (see Taplin, 2017, 19–31). The members of the "herd" – the social media users – depend on the "priests" of disruption, because through their platforms they are supplied with a continuous stream of disinformative "news" and entertainment that promises sociability and serenity:

> Facebook is filled with baby photos and cat photos because users are most likely to 'Like' baby photos and cat photos. [...] Facebook promotes fun, and users post what is fun, because on Facebook *fun* equals *increased likelihood of engagement.* (Gertz, 2024, 155; original italics)

The prime focus on engagement and thus online interaction is all about the production of data, as we have seen. The content is comparatively secondary – what matters is its relatability to an aggregate of users and their kitschy "orgies of feeling" (Anker, 2014 with a reference to Nietzsche, 1988, 92–99; 137–40; trans.), i.e., typical Internet content such as those photos, videos, conspiracy narratives, etc., that can be exploited for the benefit of the data economy. The baby photos and cat videos do not even have to be untrue to be "false." But they do enable conspiracy narratives and related technoimages to be perceived at eye level *as if* they were "true." The result is a disruption of truth, which is gradually being stripped of its foundations. In a metaphysical-ethical reading, this reevaluation means that reality and values ultimately appear as void fictions (see Lotz, 1976, 269). This may be the core narrative of post-modernism and posttruth. The world seems to be coming apart at the seams in the face of these attacks.

5 Conclusion

The persistence of posttruth narratives is remarkable. The sticky kitsch of puffed-up disinformation and conspiracy narratives thrives within them. For the producers of these informational mock-ups, the question of 'what for' – i.e., their teleological ends – is easily answered and explained by the propagandistic telos of their bullshitting. They want to keep those people that are already convinced within the grip of the discursive bubble of their self-referential beliefs, and win doubters over to their side. For the intermediaries, i.e., the tech corporations and their manifold distribution platforms, the economic purpose, as part of their business model, is their major *telos*, followed by their only slightly more subtle ideological orientation towards power and disruption (see Gertz, 2024 and Zuboff, 2019).

On the part of the users, on the other hand, there is often only a shrug of the shoulders when they encounter the posttruth of disinformation and conspiracy narratives: "[…] all we seem to be able to say is that 'Shit happens'" (Keeley, 1999, 125) – or loosely rephrasing Frankfurt (2005): *Bullshit happens*. This reflection of powerlessness is an expression of passive nihilism. Such a disposition cannot be satisfactory for a flourishing society. It may reinforce the danger that the widely multiplied contemporary falsehoods become the future gateway for even more systematic propaganda, which some states, in an intensifying "age of anger" (Mishra, 2017), as well as ideologically motivated organizations, are currently conspicuously intensifying on the international stage (on China, see Hamilton & Ohlberg, 2020, Strittmatter, 2019, Tsai, 2020, on Russia, see Burrett, 2020, Paul & Matthews 2020 and Pomerantsev 2019; on political extremism in general, see Ebner, 2017). This is being made even more complicated by the now intensified roll-out of entities of generative artificial intelligence (AI) such as refined chatbots and imaging programs that will be able to render convincing stories and deepfakes (see Risse, 2022, 98; Story & Jenkins, 2023 and Wolfram, 2023, 1–76) and virtual environments 'realistic' on levels hitherto unknown (see Chalmers, 2022). Spiteful or even just ignorant rejections and negations of the truth in the face of such onslaughts often find a fertile ground in the minds of "regressive rebels" (Amlinger & Nachtwey, 2022, 321; trans.) and may eventually deprive society of its fundament for a thriving commonality and recourses to truth. In conclusion, then, it is necessary to ask how liberal democratic societies will be able to deal with the described sticky kitschiness of disinformation, fake news, and conspiracy narratives which have been able to become so penetrating, dominant, and disruptive.

References

Amlinger, C., & Nachtwey, O. (2022). *Gekränkte Freiheit: Aspekte des libertären Autoritarismus*. Suhrkamp.

Anker, E. R. (2014). *Orgies of feeling: Melodrama and the politics of freedom*. Duke University Press.

Ball, J. (2017). *Post-truth: How bullshit conquered the world*. Biteback Publishing.
Baudrillard, J. (1978). La précession des simulacres. *Traverses, 10*, 3–37.
Beck, U. (2016). *The metamorphosis of the world*. Polity Press.
Bergmann, E., & Butter, M. (2020). Conspiracy theory and populism. In M. Butter & P. Knight (Eds.), *Routledge handbook of conspiracy theories* (pp. 330–343). Routledge.
Blackburn, S. (2005). *Truth: A guide for the perplexed*. Allen Lane/Penguin.
Bridle, J. (2018). *New dark age: Technology and the end of the future*. Verso.
Burrett, T. (2020). Evaluating Putin's propaganda performance 2000–2018: Stagecraft as statecraft. In P. Baines, N. O'Shaughnessy, & N. Snow (Eds.), *The SAGE handbook of propaganda* (pp. 492–509). SAGE.
Butter, M. (2018). *"Nichts ist, wie es scheint": Über Verschwörungstheorien*. Suhrkamp.
Butter, M., & Knight, P. (2023). Introduction: Covid-19 conspiracy theories in global perspective. In M. Butter & P. Knight (Eds.), *Covid conspiracy theory in global perspective* (pp. 3–12). Routledge. https://doi.org/10.4324/9781003330769-2
Chalmers, D. J. (2022). *Reality +: Virtual worlds and the problems of philosophy*. Penguin/Allen Lane.
Couldry, N., & Hepp, A. (2016). *The mediated construction of reality*. Polity Press.
Crosby, D. A. (1988). *The specter of the absurd: Sources and criticisms of modern nihilism*. State University of New York Press.
Ebner, J. (2017). *The rage: The vicious circle of Islamist and far-right extremism*. I.B. Tauris.
Flusser, V. (1985). Gespräch, Gerede, Kitsch. Zum Problem des unvollkommenen Informationskonsums. In H. Pross (Ed.), *Kitsch: Soziale und politische Aspekte einer Geschmacksfrage* (pp. 47–69). List.
Flusser, V. (1990). *Ins Universum der technischen Bilder* (3rd ed.). European Photography.
Flusser, V. (1996). *Kommunikologie*. Bollmann.
Frankfurt, H. G. (2005). *On bullshit*. Princeton University Press.
Frankfurt, H. G. (2006). *On truth*. Knopf.
Fuchs, C. (2023). *Digital ethics*. Routledge. https://doi.org/10.4324/9781003279488
Gadinger, F. (2019). Lügenpresse, gesunder Volkskörper, tatkräftiger Macher: Erzählformen des Populismus. In M. Müller & J. Precht (Eds.), *Narrative des Populismus: Erzählmuster und -strukturen populistischer Politik* (pp. 115–146). Springer VS. https://doi.org/10.1007/978-3-658-22374-8_7
Gelfert, H.-D. (2000). *Was ist Kitsch?* Vandenhoeck & Ruprecht.
Gertz, N. (2024). *Nihilism and technology* (Updated ed.). Rowman & Littlefield.
Gertz, N. (2019). *Nihilism*. MIT Press.
Giesz, L. (1971). *Phänomenologie des Kitsches* (2nd ed.). Fink.
Habermas, J. (2023). *A new structural transformation of the public sphere and deliberative politics*. Trans. Ciaran Cronin. Polity Press.
Hale, V. (2017). Revealed: 1,000-man mob attack police, set Germany's oldest church alight on New Year's Eve. Breitbart News., https://www.breitbart.com/europe/2017/01/03/dortmund-mob-attack-police-church-alight/
Hamilton, C., & Ohlberg, M. (2020). *Hidden hand: Exposing how the chinese communist party is reshaping the world*. Oneworld.
Harlow, R. M. (2018). Impression management. In R. L. Heath & W. Johansen (Eds.), *The international encyclopedia of strategic communication*. Wiley Blackwell. https://doi.org/10.1002/9781119010722.iesc0086
Harsin, J. (2015). Regimes of posttruth, postpolitics, and attention economies. *Communication, Culture & Critique, 8*(2), 327–333. https://doi.org/10.1111/cccr.12097
Heinrich, R. (2009). *Wahrheit*. facultas wuv.
Heller, S. (2008). *Iron fists: Branding the 20th-century totalitarian state*. Phaidon.
Hellinger, D. C. (2019). *Conspiracies and conspiracy theories in the age of Trump*. Palgrave Macmillan.
Hepp, A. (2020). *Deep mediatization*. Routledge.

Hjarvard, S. (2008). The mediatization of society: A theory of the media as agents of social and cultural change. *Nordicom Review, 29*(2), 105–134. https://www.nordicom.gu.se/sites/default/files/kapitel-pdf/269_hjarvard.pdf

Hutchinson, B., Phelps, J., and Gutman, M. (2018). Trump admits he has 'no proof' of 'Middle Easterners' in caravan, 'but there could very well be'. ABC News, October 23. https://abcnews.go.com/US/trump-admits-proof-middle-easterners-caravan/story?id=58686056

Isaacson, W. (2011). *Steve Jobs: A biography*. Simon & Schuster.

Kalpokas, I. (2020). Post-truth and the changing information environment. In P. Baines, N. O'Shaughnessy, & N. Snow (Eds.), *The SAGE handbook of propaganda* (pp. 71–84). SAGE.

Keeley, B. L. (1999). Of conspiracy theories. *The Journal of Philosophy, 96*(3), 109–126. https://doi.org/10.2307/2564659

Kelly, K. (2016). *The inevitable: Understanding the 12 technological forces that will shape our future*. Viking.

Lanier, J. (2013). *Who owns the future?* Simon & Schuster.

Lazer, D. M. J., Baum, M. A., Benkler, Y., Berinsky, A. J., Greenhill, K. M., Menczer, F., Metzger, M. J., Nyhan, B., Pennycook, G., Rothschild, D., Schudson, M., Sloman, S. A., Sunstein, C. R., Thorson, E. A., Watts, D. J., & Zittrain, J. L. (2018). The science of fake news. *Science, 359*(6380), 1094–1096. https://doi.org/10.1126/science.aao2998

Lewis, P. (2018). Steve Bannon: I want to drive a stake through the Brussels vampire. In *The Guardian*. https://www.theguardian.com/world/2018/nov/21/steve-bannon-i-want-to-drive-a-stake-through-the-brussels-vampire-populist-europe

Liessmann, K. P. (2002). *Kitsch! oder Warum der schlechte Geschmack der eigentlich gute ist*. Brandstätter.

Lotz, J. B. (1976). Nichts. In W. Brugger (Ed.), *Philosophisches Wörterbuch* (pp. 269–270). Herder.

McKenzie-McHarg, A. (2020). Conceptual history and conspiracy theory. In M. Butter & P. Knight (Eds.), *Routledge handbook of conspiracy theories* (pp. 16–27). Routledge.

Mishra, P. (2017). *Age of anger: A history of the present*. Allen Lane/Penguin.

Müller, M. (2019). Narrative, Erzählungen und Geschichten des Populismus: Versuch einer begrifflichen Differenzierung. In M. Müller & J. Precht (Eds.), *Narrative des Populismus: Erzählmuster und -strukturen populistischer Politik* (pp. 1–10). Springer VS. https://doi.org/10.1007/978-3-658-22374-8_1

Nassehi, A. (2024). *Patterns: Theory of the digital society*. Trans. Mirko Wittwar. Polity Press.

Neuberger, C. (2018). Entfesselte Kontexte: Status und Konsequenz digitaler Öffentlichkeit. In A. Nassehi (Ed.), *#realitycheck_medien (Kursbuch 195)* (pp. 31–52). Kursbuch Kulturstiftung.

Neuberger, C., Bartsch, A., Reinemann, C., Fröhlich, R., Hanitzsch, T., & Schindler, J. (2019). Der digitale Wandel der Wissensordnung: Theorierahmen für die Analyse von Wahrheit, Wissen und Rationalität in der öffentlichen Kommunikation. *Medien und Kommunikationswissenschaft, 67*(2), 167–186. https://doi.org/10.5771/1615-634X-2019-2-167

Nietzsche, F. (1988). *Zur Genealogie der Moral: Eine Streitschrift*. Reclam. [first published 1887].

Nietzsche, F. (1994). *Also sprach Zarathustra: Ein Buch für Alle und Keinen*. Reclam. [first published 1883–85].

Nocun, K., & Lamberty, P. (2020). *Fake Facts: Wie Verschwörungstheorien unser Denken bestimmen*. Quadriga.

O'Shaughnessy, N. (2020). From disinformation to fake news: Forwards into the past. In P. Baines, N. O'Shaughnessy, & N. Snow (Eds.), *The SAGE handbook of propaganda* (pp. 55–70). SAGE.

O'Toole, F. (2018). The paranoid fantasy behind Brexit: In the dark imagination of English reactionaries, Britain is always a defeated nation – and the EU is the imaginary invader. In *The Guardian*. https://www.theguardian.com/politics/2018/nov/16/brexit-paranoid-fantasy-fintan-otoole

Paul, C., & Matthews, M. (2020). Defending against Russian propaganda. In P. Baines, N. O'Shaughnessy, & N. Snow (Eds.), *The SAGE handbook of propaganda* (pp. 286–302). SAGE.

Polizei NRW Dortmund. 2017a. POL-DO: Amtliche Fakten der Polizei zur Silvesternacht 2016/17. Lfd. Nr.: 0015. Presseportal.de., https://www.presseportal.de/blaulicht/pm/4971/3527791

Polizei NRW Dortmund. (2017b). Amtliche Fakten der Polizei zur Silvesternacht 2016/17. In *Twitter.com*. https://twitter.com/polizei_nrw_do/status/817770414084591616?s=20

Pomerantsev, P. (2019). *This is not propaganda: Adventures in the war against reality*. PublicAffairs.

Precht, J. (2019). Verbreitung populistischer Narrative in den Kommentarspalten des sozialen Netzwerks Facebook. In M. Müller & J. Precht (Eds.), *Narrative des Populismus: Erzählmuster und -strukturen populistischer Politik* (pp. 93–114). Springer VS. https://doi.org/10.1007/978-3-658-22374-8_6

Raffnsøe, S. (2007). *Nietzsches "Genealogie der Moral"*. Fink.

Risse, M. (2022). Artificial intelligence and the past, present, and future of democracy. In S. Voeneky, P. Kellmeyer, O. Mueller, & W. Burgard (Eds.), *The Cambridge handbook of responsible artificial intelligence: Interdisciplinary perspectives* (pp. 85–103). Cambridge University Press. https://doi.org/10.1017/9781009207898.009

Sotirovic, M. (2020). Trumpaganda: The war on facts, press, and democracy. In P. Baines, N. O'Shaughnessy, & N. Snow (Eds.), *The SAGE handbook of propaganda* (pp. 510–531). SAGE.

Stalder, F. (2018). *The digital condition*. Trans. Valentine A. Pakis. Polity Press.

Stark, B., Magin, M., & Jürgens, P. (2018). Politische Meinungsbildung im Netz: Die Rolle der Informationsintermediäre. *UFITA. Archiv für Medienrecht und Medienwissenschaft, 82*(1), 103–130. https://doi.org/10.5771/2568-9185-2018-1-103

Story, D., & Jenkins, R. (2023). Deepfake pornography and the ethics of non-veridical representations. *Philosophy and Technology, 36*, 56. https://doi.org/10.1007/s13347-023-00657-0

Strittmatter, K. (2019). *We have been harmonised: Life in China's surveillance state*. Trans. Ruth Martin. Old Street Publishing.

Sunstein, C. R., & Vermeule, A. (2009). Symposium on conspiracy theories: Conspiracy theories: Causes and cures. *The Journal of Political Philosophy, 17*(2), 202–227. https://doi.org/10.1111/j.1467-9760.2008.00325.x

Taplin, J. (2017). *Move fast and break things: How Facebook, Google and Amazon have cornered culture and what it means for all of us*. Macmillan.

Tolentino, Daysia. (2023). AI-generated images of Pope Francis in puffer jacket fool the internet: The viral images are one of the first instances of wide-scale misinformation stemming from artificial intelligence. NBC News., https://www.nbcnews.com/tech/pope-francis-ai-generated-images-fool-internet-rcna76838

Tsai, C.-M. (2020). The construction of the chinese dream. In P. Baines, N. O'Shaughnessy, & N. Snow (Eds.), *The SAGE handbook of propaganda* (pp. 405–421). SAGE.

Vuger, D. (2021). The technical image in Vilém Flusser. In K. Purgar (Ed.), *The Palgrave handbook of image studies* (pp. 379–393). Palgrave Macmillan. https://doi.org/10.1007/978-3-030-71830-5_23

Waldvogel, T. (2018). Die postmoderne Wahrheit vs. die Wahrheit_Wahrheit. In A. Bieber & F. Waldvogel (Eds.), *Im Zweifel für den Zweifel: Die große Weltverschwörung* (pp. 25–34). NRW-Forum.

Wolfram, S. (2023). *What is ChatGPT doing … and why does it work?* Wolfram Media.

Zöllner, O. (2019). Das Picknick von Bucklebury: ein Genrebild erfolgreicher Herrschaftsdiskurse im Medienzeitalter: Eine dokumentarische Analyse. In M. Junge (Ed.), *Das Bild in der Metapher: Bilder des Erfolgs – Bilder des Scheiterns* (pp. 89–115). Springer VS. https://doi.org/10.1007/978-3-658-24562-7_7

Zuboff, S. (2019). *The age of surveillance capitalism: The fight for the future at the new frontier of power*. Profile Books.

Explanations and Counter-Measures

Ignorance in Social Networks: Discounting Delays and Shape Matters

Brian Ball, Alexandros Koliousis, Amil Mohanan, and Mike Peacey

Contents

Abstract Knowledge is the proper basis for action (Williamson, Timothy, *Knowledge and its limits*. Oxford University Press, Oxford, 2000). But disinformation causes ignorance, whether through error (false belief) or omission (agnosticism). We use philosophical simulations (Mayo-Wilson, Conor and Kevin Zollman, *Synthese* 199:3647-3673, 2021) to study how ignorance persists in networks of inquiring rational agents. Following (Zollman, Kevin J. S., *Philosophy of Science* 74:574-587, 2007), we simulate communities of agents who generate evidence, share it with their neighbours, and then update their beliefs. After recapping previous results, we report two novel findings. First, in variations on the mistrust models developed by O'Connor and Weatherall (O'Connor and Weatherall, *European Journal for Philosophy of Science* 8:855-875, 2018), discounting evidence received from neighbours delays convergence to the truth. Second, ignorance persists differentially in networks of different shapes, even when they have the same overall connectivity. These results shed light on the structural causes of ignorance that can be exploited by those engaged in disinformation campaigns; and they constrain the space of knowledge-conducive responses.

B. Ball (✉) · A. Koliousis · A. Mohanan
Northeastern University London, London, UK
e-mail: brian.ball@nulondon.ac.uk; alexandros.koliousis@nchlondon.ac.uk; amil.mohanan@nulondon.ac.uk

M. Peacey
University of Bristol, Bristol, UK
e-mail: mike.peacey@bristol.ac.uk

M. M. Resch et al. (eds.), *Trust and Disinformation*,
https://doi.org/10.1007/978-3-031-96790-0_7

1 Introduction

Actions undertaken by social groups are best when underpinned by relevant knowledge. Appropriate responses to climate change, or the coronavirus pandemic, for example, require informed (coordinated) democratic action on the part of whole societies. And these actions may be legitimated by the inquiries of e.g. certain scientific communities. Yet mis- or even disinformation may disrupt knowledge from taking hold within groups of individuals, preventing timely and appropriate action. Indeed, knowledge is hard won, and ignorance may persist or prevail even in their absence.

The current paper investigates ignorance in social networks. We begin by describing our approach, detailing key aspects of our models and methods. We then summarize previous results (surrounding the Zollman effect and polarization) before presenting two novel findings of our own: first, that discounting good evidence delays the emergence of a correct, knowledgeable consensus within communities of rational inquirers; and second, that when it comes to the persistence of ignorance, social network shape, or topology, matters. We close with some concluding remarks that contextualize our findings, indicating their relevance to issues of trust and disinformation.

2 Models and Methods

We model the social situations that interest us as involving networks of individual rational agents, conducting experiments, sharing their findings, and updating their beliefs in light of the evidence at their disposal (cf. Zollman, 2007). Allow us to unpack this.

2.1 Social Groups as Networks of Individuals

We model social groups as networks of individuals. These networks can be represented as *graphs*: (sets of) *nodes* connected by *edges*. Although the terminology here derives from network theory, the approach should be familiar to those acquainted with (the semantics of) predicate logic: a given graph represents the extension of a relation amongst (or, on a domain consisting of) some individuals. Indeed, multi-edge graphs are possible – these represent the extensions of a number of different relations simultaneously (on the domain in question). Moreover, labels can be attached to the nodes: this allows for the representation of properties of the individuals. Here we focus on the simple case of unlabelled, single-edge graphs.

Note that, while it is common to present the semantics of predicate logic – and network theory – in the language of set theory, so that domains are sets of individuals, this is not essential: a plural framework can be employed instead. We will not make much of this here: but it is worth remarking that we are not assuming that social groups are *sets* of individuals; for sets are themselves individual entities over and above, and distinct from, their members. (For example, even the singleton set {i} comprising just the individual i is distinct from i itself.) We are rather concerned with e.g. scientific communities, or democratically self-governing peoples, construed as the individuals themselves – taken collectively.[1]

Indeed, there has been much interesting work recently on the metaphysics of social groups. For instance, (Uzquiano 2004) argued that the Supreme Court of the United States must be construed as a *sui generis* entity distinct from its justices (yet not a set or mereological sum of them either). While we are sympathetic to there being such entities (e.g. corporations, and other institutional organizations), here we more closely follow the work of (Ritchie, 2020), for whom "social structures are central to the nature of all social groups" (ibid., 402), and where structures in turn are "complexes… of relations" (ibid., 405). Thus, we crucially take the social groups with which we are concerned, to be not mere pluralities (or sets!) of individuals, but *structured* pluralities, or *networks* of individuals – which can be appropriately represented by graphs.

2.2 *Individuals as Rational Agents*

At the nodes of our networks are rational agents. As *agents*, they take actions. These actions in turn modify their environment; thus, the effects of their actions can be observed, yielding new evidence. In our models, the actions are limited to two: agents perform either action A or action B. In particular, we can think of the agents as (scientists) conducting experiments, such as trialling a drug on a certain number *n* of experimental patients.

Insofar as they are *rational*, our agents' beliefs are based on the evidence at their disposal, and the actions they take are appropriate given their attitudes. Thus, they perform action A (administer a familiar drug) if they take it to be more likely to yield a desirable result (e.g. help more of their experimental patients), and B if they think (as is in fact the case) B is better (in this same sense).[2] They observe the results

[1] It is common to distinguish 'collective' from 'distributive' plural predications: the latter are equivalent to conjunctions of individual predications, while the former are not; thus, the claim that some people carried a piano is likely meant collectively – unless the people in question are e.g. participants in a strongman competition.

[2] More sophisticated models of the practical rationality involved in deciding what action to take are of course possible. Here it is the agents' beliefs alone that determine what they do; but we might expect their desires, preferences, or utilities to play a role as well - as in standard decision theory. For instance, O'Connor & Weatherall (2019) discuss 'conformity' models in which scientific agents are motivated not only by the desire to ascertain the truth about the matter they are investigating, but also by a desire to conform with what their peers are doing. We do not here explore such variations on the basic models under discussion.

of their experiments (e.g. six out of ten patients recovered), and communicate their findings (by testimony) to the people with whom they are connected (by edges) in the network – their *neighbours*. Finally, they update their beliefs in light of the evidence at their disposal (that which they themselves generate, as well as that which is communicated to them by their neighbours).

What exactly is rational belief updating? We consider two variants. In both, our agents are modelled as having a *credence*, or degree of belief (between 0 and 1), in the (true) proposition that action (or drug) *B is better* (than A).[3] A is assumed to have a constant (and known), 0.5 probability (i.e. 50% chance) of helping a given patient to recover from their illness; B has a 0.5 + (positive) epsilon probability of doing so. The proposition that B is better just is the proposition that epsilon is positive – which in our models it always is (so that our target proposition is true); but this is not initially known to the agents in the models.

In the models originally employed by Zollman (2007), agents are initially assigned a credence drawn from a random uniform distribution (in line with the *subjective* Bayesian assumption that any credence is rational prior to the receipt of evidence). They then update their beliefs in light of the evidence available at any given stage using Bayes' rule:

$$P_f(h) = P_i(h|e) = P_i(e|h) / P_i(e) \tag{1}$$

(where P_f is the final probability, after updating, and P_i the initial probability, prior to updating; *h* is some hypothesis, and *e* is a given piece of evidence).

Indeed, O'Connor & Weatherall, (2019) argue (based on diachronic Dutch book considerations) that this is the unique rational method for updating beliefs over time.[4] Nevertheless, they themselves subsequently suggest that it might sometimes be rational to be more cautious in dealing with the available evidence. Thus, in their variant 'mistrust' models, agents are uncertain of the evidence with which they are presented, and update using Jeffrey's rule:

$$P_f(h) = P_f(e) \times P_i(h|e) + P_f(\neg e) \times P_i(h|\neg e). \tag{2}$$

It is worth noting that Bayes' rule is the special case of Jeffrey's rule in which $P_f(e) = 1$ and $P_f(\neg e) = 0$. Thus, the use of Jeffrey's rule might be thought to simply generalize the use of Bayes' rule. However, the matter is not quite so simple: as Williamson, (2000) has noted, Jeffrey's rule cannot be immediately operationalized; in order to put it to use, we need to know what $P_f(e)$ is – and nothing in the rule itself determines this.

---

[3] Note that this is already a robust theoretical assumption: some theorists regard (outright) belief as a basic, all-or-nothing affair; that belief comes in degrees, as assumed in the models we employ, is therefore contentious.

[4] Such arguments are contentious – see Williamson (2000).

In order to address this issue, and put Jeffrey's rule into practice in their models, O'Connor and Weatherall make an assumption of *homophily*: people trust others more when they are more alike.[5] Moreover, they assume a particular version of this idea – namely, that people trust each other more when their *beliefs* (on the question whether B) are more similar.[6] Finally, they implement this (attitudinal) version of the assumption in a *specific mathematical formula*[7]:

$$P_f(e) = 1 - \min(\{1, d \times m\}) \times (1 - P_i(e)). \tag{3}$$

Here d is the distance (i.e. the absolute value of the difference) between the credences of the recipient and the provider of the evidence supplied through testimony. Thus, agents do not discount the evidence they themselves produce at all, since $d = 0$: accordingly, the final probability of the evidence they provide themselves with is 1; that is, they treat this evidence as certain. But for evidence supplied by other agents, as d increases, the product of d and the 'mistrust multiplier' m increases until it reaches (and then exceeds, but is replaced by the smaller value) 1: at this point $P_f(e) = P_i(e)$; in other words, the evidence provided is simply ignored (and treated as precisely as certain as it was prior to its being given).

2.3 *The Method of Philosophical Simulation*

Like the researchers discussed above, we run computer simulations based on our models of the social situations we wish to investigate. Our (Python) code first builds a network (of a specified kind and size – see below) and assigns initial

---

[5] Note that it is possible to implement Jeffrey's rule without the assumption of homophily. For instance, in our 'testimonials' models (Ball et al., 2024b), we consider the case in which agents set the final probability of the evidence they receive to the frequency of truth-telling in their network and/or neighbourhood.

[6] It is also possible to implement the homophily assumption without assuming that the relevant respect of similarity is in the credence given to the proposition that B is better. For instance, in our 'epistemic injustice' models (Ball et al., 2024b), we label nodes to indicate different social groups. We might then set one (relatively low, possibly null) discount rate for evidence received from those in the same group, and one or more other (higher) rates for those from different groups. This would embody a homophily assumption not based on differences in credences (which, after all, might be difficult to determine). An alternative approach would simply set a single discount rate applied by all agents determined by the group label of the speaker, thereby abandoning the homophily assumption entirely.

[7] In fact, O'Connor & Weatherall, (2018) use two different rules: they also run simulations using a rule - namely, $P_f(e) = \max(\{1 - d \times m \times (1 - P_i(e)), 0\})$ - that allows for 'anti-updating' (whereby the final credence in e is less than the initial credence in that same proposition). Here we discuss only simulations with no anti-updating - i.e. those using the rule given in the main text.

credences to the agents at its nodes (drawn from a random uniform distribution in all of our simulations to date). It then simulates that group of agents conducting experiments, sharing results, and updating credences in the manner specified in the model – with the values of certain key parameters (e.g. epsilon, number of trials) set. It runs until all agents have credence less than 0.5 or over 0.99 in B – or else, in some mistrust models (those leading to polarization – see below), the differences in credences preclude any further dynamics (because any new evidence produced by B believers is ignored – i.e. completely discounted – by the A believers).

Some might worry that our simulations will not be appropriately realistic – that is, that they may not have 'external validity'. In particular, it might be thought that the simulations we run make erroneous assumptions about either: (i) the nature of the individuals that constitute our social groups (specifically, it might be thought that people are not rational though we assume they are); or (ii) the character of the networks that bind them into communities; or both. If that is right, then (it is argued) we will not be able to draw (legitimate) conclusions – and in this way learn – about the situations we are modelling from findings about our simulations.

In response to this objection, and by way of clarification of our method, we must, unfortunately, be brief. We begin with point (i). First, while many irrationalities have been suggested (Kahneman, 2011), there is some debate around whether reliance on such heuristics is genuinely irrational (Mousavi & Gigerenzer, 2017). Second, we are idealizing: just as physicists model situations as involving no air resistance, or a frictionless plane, so we too imagine our social groups as comprising agents without irrationalities. This simplification allows us to approximate the truth before refining our models. Third, relatedly, this allows us to identify causes and provide explanations. Our aim is not (yet, at least) to make predictions about how social groups will behave when investigating a given issue under particular circumstances. (Thus, in a sense we require less of the external validity of our simulations than some might expect.) More specifically, fourth, we wish to uncover *structural* causes of ignorance (e.g. due to misinformation): if we assume (even if falsely) that agents are perfectly rational, and we nevertheless discover that communities of these agents remain ignorant of the truth in our simulations (for an extended period of time), then it will be apparent that (at least some) ignorance has a cause other than the psychological failings of the agents constituting them. These causes ultimately derive from the (social) structures within which the agents are embedded. And finally, fifth, we are also engaged in an interpretive, and normative, (humanities) investigation (not just a descriptive, scientific one): we are aiming to assess (a) whether our models of rational agents are appropriate, or *apt*, in the sense that they capture what we have in mind when we think of (the ideal of) rationality (this is the interpretive element), and (b) whether behaving in the ways our agents do really is rational (especially in light of the foreseeable social consequences) – that is, whether agents really *ought* to behave in those ways (this is the normative aspect). For all of these reasons, we believe that much can be learned about real social situations, even through simulations based on somewhat simplified models.[8]

Turning to (ii), we are aware of the concern – which is why, in addition to looking at highly controlled, artificial and relatively small networks, we also run our simulations on networks drawn from real world data sets (Ball et al., 2024a). Thus, unlike (at least some) previous efforts in 'computational philosophy' (Grim & Singer, 2022), our framework allows us to *scale* our simulations to run on large data sets – and we exploit the opportunities this affords. Moreover, whatever our findings in relation to smaller simulations on artificially generated networks, we hope to extrapolate from them to explain findings based on real world data sets with the structural features, whatever they might be, that actual social networks possess. For instance, we may find cleaner patterns within more controlled environments that may nevertheless be discerned – albeit with some noise – in more complex, realistic settings; though they might have been difficult to detect, or recognize, if we worked exclusively with real-world data sets. Nevertheless, we here focus on simulations run on small, artificial networks.

3 Results

Researchers employing philosophical simulations built on the models described above have obtained a number of results. We begin by sketching previous results, before reporting our own findings.

3.1 The Zollman Effect

Zollman, (2007) found (amongst other things) that simulations (using Bayesian updating) converged to the true belief that B is better more quickly (i.e. in fewer steps) in complete networks (see Fig. 1) in which each agent is connected (by an

[8] Mayo-Wilson and Zollman defend 'simulation as a core philosophical method' (2021, 3647), offering two central arguments, and responding to objections. The second of their arguments is irrelevant for our purposes (since it concerns benefits to the researchers, rather than the research), but the first depends on the crucial claim that computer simulations can play many of the same roles as thought experiments do – indeed, they argue that the former can be better at doing so in relation to five of the six aims they identify for the latter. We are inclined to agree with much of what they say, and note that one of the aims they mention is to '[d]istinguish explanatory reasons and identify those causes that explain a phenomenon' (2021, 3650) – which is, at least roughly, one of the objectives we have given above in relation to our simulations. The only aim that Mayo-Wilson and Zollman recognize in the case of thought experiments, but do not argue is well served by simulations is that of '[e]licit[ing] normative intuitions' (2021, 3650) – but that is something that we do hope our simulations can help us to do, as hinted above. Of course, we agree that 'simulations almost never answer philosophical questions by themselves' (2021, 3649) – but we think they can play a role, within a given dialectical context, in drawing out normative intuitions that can help to settle an issue (such as what rationality requires).

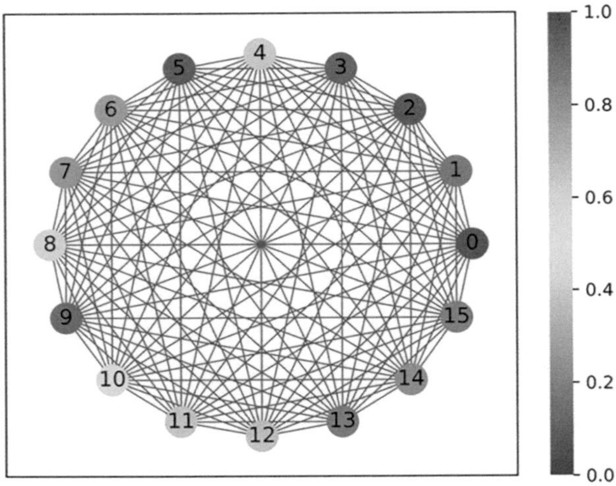

Fig. 1 Complete network (size 16)

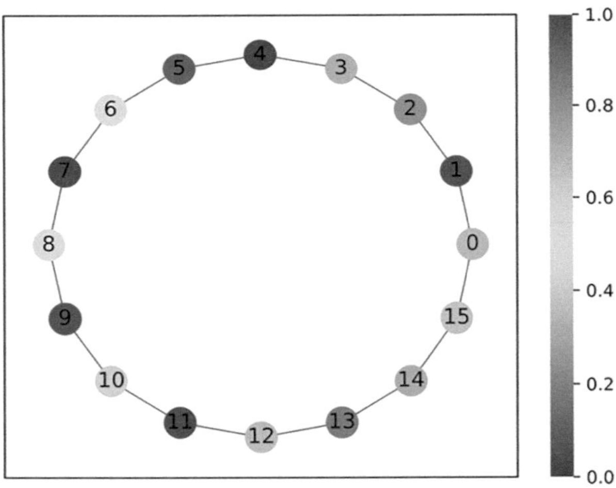

Fig. 2 Cycle network (size 16)

edge) to every other than in cycle networks (see Fig. 2) where each agent is connected to exactly two neighbours (with the network as a whole forming a ring).[9] But he also found that the complete networks were more likely than cycles to converge to the false belief that A is better. In his own words, his results suggested "two interesting conclusions. First, in some contexts, a community of scientists is, as a whole, more reliable when its members are less aware of their colleagues' experimental results. Second, there is a robust tradeoff between the reliability of a community and the speed with which it reaches a correct conclusion" (2007, 574).

[9]Zollman looked at various other small networks as well. Some of these will be discussed below.

O'Connor and Weatherall introduced the term 'Zollman effect' as a name for the 'phenomenon' in which "scientists improve their beliefs by failing to communicate" (2019, 63) – i.e. the first of Zollman's findings (in his own report above). Interestingly, however, Rosenstock et al., (2017) speak of "the general effect noted by Zollman—that sparser networks are more reliable than well-connected ones". Now, one network is sparser than another, when it has lower *density* – where this is the number of actual connections (or edges) in the network divided by the number of potential connections (which itself is half the number of nodes n multiplied by n-1). (Thus, a complete network has the highest possible density: 1.) And yet O'Connor & Weatherall, (2018) suggest (in footnote 28 on page 198) that the Zollman effect concerns 'social structure': and while density is determined by structure, there is more to the latter than the former. We will return to this below.

In any case, Rosenstock, Bruner, and O'Connor report "results replicating Zollman's simulations with a wider parameter space" (Rosenstock et al. 2017, 241) than Zollman himself employed. They "find that parameters for which there is a notable benefit to decreased network connectivity occupy a relatively small niche of the total space" (2017, 241). In particular: "the difference in likelihood of successful convergence between the two network configurations [cycle and complete] decreases" (2017, 241) as epsilon increases; "the difference in rates of successful convergence for the cycle and complete networks is more significant for lower [number] *n* [of trials]" (2017, 243); and crucially, "[t]he Zollman effect is strongest for smaller networks, but as network size increases it drops off" (2017, 244).

3.2 *Polarization*

O'Connor & Weatherall, (2018) ran simulations in complete networks using Jeffrey's rule for updating, with the evidence provided by others being discounted increasingly with larger differences in credences (as described above). (As we have seen, this assumes homophily: we trust others more when they are more like us – and in particular, when their opinions are similar to our own.) They found that their simulations often ended in a state they refer to as 'polarization'. Of course, polarization can be – and has been – understood in a number of ways. O'Connor and Weatherall themselves say that "polarization involves the emergence of two subgroups, one whose members all have credence > .99, and the rest with a variety of stable, low credences, such that they prefer the worse theory [that A is better]" (2018, 866).[10]

Reporting their findings on polarization (so understood), O'Connor and Weatherall say: "In our models,... over all parameter values, we found that only 10% of trials ended in false consensus, 40% in true consensus, and 50% in

[10] They go on to say: "(More precisely, a stable outcome is one in which every agent either (a) has credence > .99 or else (b) has credence <= .5 such that their distance to all agents whose credence is > .99 satisfies m $*$ d > = 1.) Because the agents with low credences are outside the "realm of influence" of those testing the informative theory, they do not update their beliefs" (2018, 866).

polarization. These values should not be taken too seriously, since parameter choices influence where and when polarization happens, but the point is that adding evidential assessments based on shared belief dependably generates stable polarization." (2018, 866) This requires unpacking, as well as qualification.

We begin with the qualification. In the (no anti-updating) formula (given above) that O'Connor and Weatherall (2018) use to operationalize Jeffrey's rule in the context of their homophily assumption, polarization is only a possible outcome when the mistrust multiplier is sufficiently large that min({1, $d \times m$) is 1. Since d itself is a real number between 0 and 1, this is only the case when $m > 1$. Thus, adding evidential assessments based on shared belief dependably generates stable polarization in this case only. So much for the qualification; now for the unpacking. Unfortunately, O'Connor and Weatherall do not themselves provide the information needed to unpack their findings in the case where mistrust can lead to polarization. Elsewhere (Ball et al., 2024a), we report the results of the simulations we ran (on complete networks) to uncover the requisite details. In brief, we found that ignorance – whether through error (i.e. false belief) or omission (i.e. agnosticism due to polarization) – was more likely to prevail in the long-run in O'Connor and Weatherall's polarization models (with $m > 1$) than in Zollman's original models (keeping other parameter settings fixed). Here we focus on the case where the mistrust multiplier $m \leq 1$.

3.3 Discounts Delay

What happens in O'Connor and Weatherall-style models when polarization is not a possible outcome? To find out, we ran simulations on complete networks, using O'Connor and Weatherall's (no anti-updating) operationalization of Jeffrey's rule with mistrust $m = 1$. Since polarization is not a possible outcome in this case, we did not see the increased ignorance in the long-run due to agnosticism that we saw for $m > 1$. But when it comes to some of the issues that must be resolved through collective decision-making, such as efforts within democracies to avoid catastrophic climate change, timeliness is important; and so we looked at how long ignorance persisted in these models when it was eventually overcome. More specifically, we compared the mean number of steps in simulations that converged on the correct consensus that B is better in these models and in Zollman's original models. To ensure the robustness of our findings, we also considered ways of amplifying or dampening the effect of the distance between credences that, unlike in O'Connor and Weatherall's rule, cannot result in polarization. Thus, rather than multiplying the initial probability of the negation of the evidence $(1 - P_i(e))$ by $min(\{1, d \times m\})$ before setting the final probability of the evidence as 1 minus the result, we explored the effects of multiplying by the square, and by the square root, of d instead.[11] Table 1 below shows our findings.

[11] When $m = 1$, O'Connor and Weatherall's no anti-updating rule is equivalent to $P_f(e) = 1 - d \times (1 - P_i(e))$. We also considered the d-squared rule, $P_f(e) = 1 - d^2 \times (1 - P_i(e))$, and the root- d rule, $P(f) = 1 - \sqrt{d} \times (1 - P_i(e))$. Since $d \leq 1$, squaring dampens the effect of distance in discounting the evidence, while taking the root amplifies it.

Table 1 Comparing O'Connor and Weatherall-style simulations with Zollman's original models. Jeffrey's rule is employed and evidence is discounted in line with a homophily assumption in the former. In the latter, Bayes' rule is used, and evidence is fully trusted. We used a Mann-Whitney U-test ($p < 0.05$) to show that ignorance persists for longer (i.e. more steps) in homophily models, even when it is eventually overcome. Discounting good evidence delays convergence to the truth

Size	Epsilon	Trials	Model	Total (count)	Steps (mean)	U value	P value
16	0.001	16	Zollman	558	5121		
			Square(d)	94	7867	36,844.5	0.00
			Mistrust m=1	430	14,649	193,902	0.00
			Root(d)	91	19,215	47,956	0.00
	0.001	64	Zollman	577	1287		
			Square(d)	95	1955	38,336	0.00
			Mistrust m=1	421	3233	202,012	0.00
			Root(d)	98	7499	52,553	0.00
	0.01	16	Zollman	93	53		
			Square(d)	95	78	5506	0.00
			Mistrust m=1	430	129	31,830.5	0.00
			Root(d)	96	228	8306	0.00
	0.01	64	Zollman	99	14		
			Square(d)	95	19	6103	0.00
			Mistrust m=1	435	59	34,939	0.00
			Root(d)	99	9962	8908.5	0.00
64	0.001	16	Zollman	597	1710		
			Square(d)				
			Mistrust m=1	451	7197	234,197	0.00
			Root(d)	100	15,738	57,935.5	0.00
	0.001	64	Zollman	595	419		
			Square(d)	97	688	41,121.5	0.00
			Mistrust m=1	456	1575	244,823.5	0.00
			Root(d)	97	2987	56,180.5	0.00
	0.01	16	Zollman	98	16		
			Square(d)	98	31	7251.5	0.00
			Mistrust m=1	452	61	39,782.5	0.00
			Root(d)	99	89	9.475	0.00
	0.01	64	Zollman	100	5		
			Square(d)	98	6	6292.5	0.00
			Mistrust m=1	453	13	38,155.5	0.00
			Root(d)	98	22	9326.5	0.00

As can be seen, agents' discounting of the evidence they receive in these models leads to significant delays in arriving at the correct consensus that B is better. Moreover, the mean number of steps required to converge to the truth increases as the discounting effect of the distance between credences becomes more pronounced. It should, of course, be remembered that, in these models, all agents are reliable truth-tellers, reporting the results of their experiments accurately, so that the evidence received is factive (i.e. true), even if (due to its probabilistic nature) potentially misleading as regards the underlying issue under investigation. It is therefore possible that in other contexts, where information sources are less reliable, discounting what is presented as 'evidence' may yield benefits. Nevertheless, the current findings provide a kind of benchmark result: discounting good evidence delays convergence to the truth; ignorance persists longer under mistrust than full trust.

3.4 Shape Matters

We reported (above) the finding, due to Zollman, that sparser (small) networks converge to the truth more reliably than denser ones: that is, for a given size of (small) network, the proportion of simulations that achieve consensus in the long run that B is better than A decreases with the density of the network; or, in other words, better connected (small) networks are more error prone.

For a given network size, the density of the network determines the average neighbourhood size in that network. Moreover, how much evidence a node receives is controlled by (the number of trials conducted by each agent and) the size of its neighbourhood. (It is not strictly *determined* by this, since how many of its neighbours take action B can vary while the size of its neighbourhood does not.) So network size and density together control the quantity of evidence each node receives on average at any given simulation step. Thus, what Zollman found was that, for a given size of (small) network, there being more information on average for each node yields a consensus more quickly, but – somewhat surprisingly – with a greater chance of error.

But there is more to network structure than density alone. Each network has a specific shape (or topology). Above we saw the complete (Fig. 1) and cycle (Fig. 2) networks. (Clearly, the complete network is more densely connected than the cycle.) A star network (Fig. 3) consists of a central node with all other nodes connected only to (themselves and) it. Networks of other kinds are generated by stochastic processes, with different instances of the kind having somewhat different shapes. For instance, a Barabasi-Albert network (Fig. 4) is generated by 'preferential attachment': as each new node is added to the network it forms a fixed number of attachments to previous nodes (1 in the network shown in Fig. 4), with the probability of forming an edge connecting it to a given node determined by the size of the neighbourhood of that other node; in other words, the more edges a node has, the more likely it is for the next new node to attach to it. Random networks (not pictured) are formed by making edges between any two nodes with a given, fixed probability; and Watts-Strogatz networks (not pictured) begin as rings, or cycles, of a given thicknesses (with each node connected to e.g. 2 neighbours, or 4, or…), with edges then rewired, with a certain probability, to connect to a randomly chosen node.

Do other structural features of networks, besides density, influence the knowledge or ignorance of the group? We compared simulations run (using Bayes' rule) on networks of different kinds, but with similar densities, and found that, even when

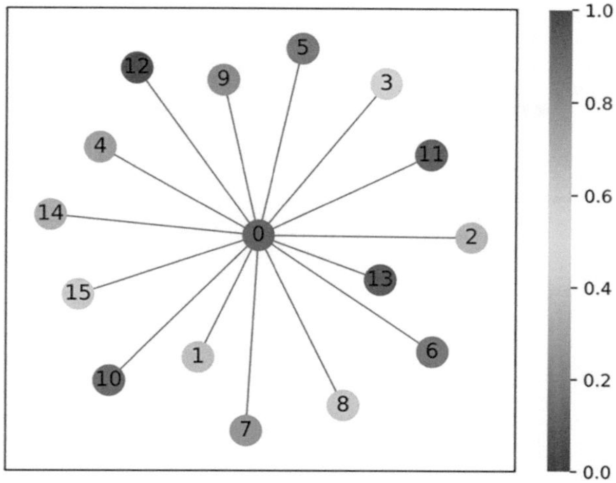

Fig. 3 Star network (size 16)

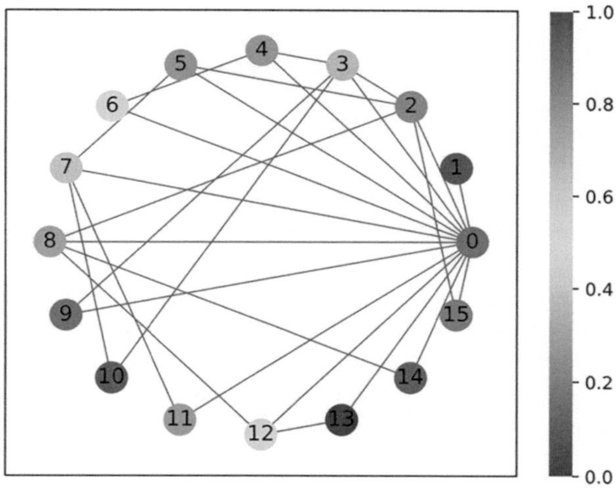

Fig. 4 Barabasi-Albert network (size 16)

it was eventually eradicated (with all the nodes converging on the correct consensus that B is better), ignorance persisted significantly longer in networks of some kinds as opposed to others. For example, in networks of size 64, cycles and stars, as well as certain Barabasi-Albert (attachments: 1) and Watts-Strogatz (knn: 2) networks, the density of connections is approximately 0.03 (i.e. roughly 3% of possible connections are actualized). And yet (with epsilon 0.001) the number of steps required for convergence to B differed significantly between the simulations run on these networks. Stars were the fastest. Cycles were next, though much slower. Barabasi-Albert networks slower still; and Watts-Strogatz networks were the slowest. We

Table 2 Comparing the number of steps required for convergence to B in various kinds of (size 64) networks with similar densities (all approximately 0.03). In these simulations, each B node conducted 64 trials, and the value of epsilon was 0.001. Here we report the number of simulations of each kind, as well as the U- and p-values (<0.05) for each Mann-Whitney U-test performed

	Star (mean steps: 68,818)	Cycle (mean steps: 109,127)	Barabasi-Albert (mean steps: 119, 976)	Watts-Strogatz (mean steps: 120, 596)
Star (n = 600)	–	U = 294,660.5 p = 0.00	U = 43,898.5 p = 0.00	U = 31282.0 p = 0.00
Cycle (n = 600)		–	U = 134,036.0 p = 0.00	U = 98,518.5 p = 0.00
Barabasi-Albert (n = 500)			–	U = 107,933.5 p = 0.04
Watts-Strogatz (n = 400)				–

tested each pairwise comparison for significance ($p < 0.005$) using a Mann-Whitney U-test – our findings are in Table 2 below.[12]

These considerations show that the persistence of ignorance in a social network of rational (Bayesian) agents is influenced not only by the density of connections in that network, but also by the kind of network it is: the number of steps needed for convergence to the truth is significantly different between networks of different kinds with similar densities.

Interestingly, Zollman claimed that, in the small networks he tested, "the in-network degree variance is not correlated with success" (Zollman, 2007, 482): that is, while the average neighbourhood size within a network affects reliability, variation in neighbourhood size does not. However, we found that variation in the size of the nodes' neighbourhoods can make a difference to the persistence of ignorance. We used a normalized measure we called 'moment_sd': it is the standard deviation in neighbourhood size divided by the mean neighbourhood size. Figure 5 (below) shows that (with epsilon 0.001 and 16 trials) steps increase as this measure increases in Barabasi-Albert and Watts-Strogatz networks with density 0.53.[13] While more work is needed to understand the way(s) in which it does so, it is clear that network shape matters.

[12] With the parameter setting above, but with 16 (rather than 64) trials, we again found significant differences between networks kinds, except between Barabasi-Albert and Watts-Strogatz; the same was true for size 16 networks of these kinds, with both 16 and 64 trials. Increasing the numbers of attachments and nearest neighbours, however, we did see significant differences between Barabasi-Albert and Watts-Strogatz networks with both 16 and 64 trials, for: size 16, density 0.53; size 64, density 0.06; and size 64, density 0.12.

[13] The trend seen here was less pronounced, but still present, in these networks when trials increased to 64; and in size 64 networks with densities 0.12 and 0.06. In the lowest density networks (i.e. Barabasi-Alberts with 1 attachment and Watts-Strogatz, knn: 2) there was no significant correlation.

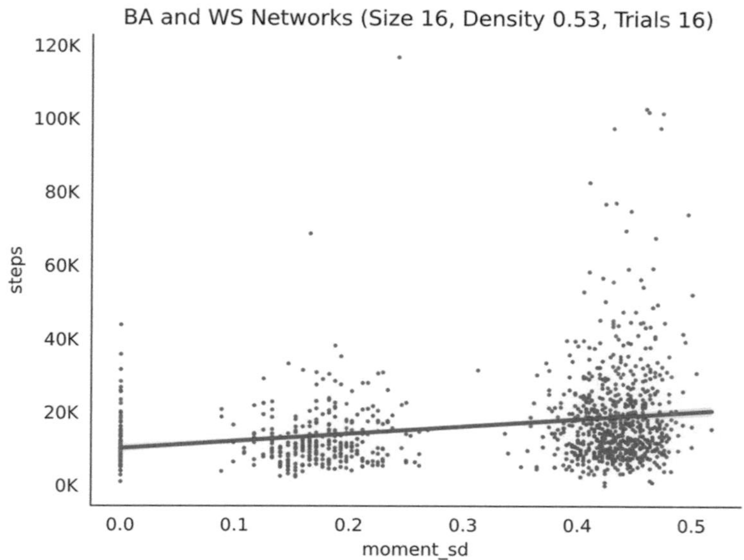

Fig. 5 The effect of moment_sd on steps. R-squared: 0.064014. Slope (CI: 95%): 20513.379535 +/− 4644.725867. Number of simulations: 1100

4 Concluding Remarks

We have explored ignorance in social networks, discussing computer simulations that employ simple models of inquiring communities. We avoided imbuing the agents in these models with psychologically motivated departures from rationality, in the hope of uncovering other (social and environmental) reasons why knowledge may fail to emerge in a timely fashion. Building on previous work, our results show: first, that when all agents truthfully report their own observations, discounting the evidence they provide delays any eventual correct consensus; and second, that the shape (and not just the connectivity) of the network formed by a community of agents influences how long ignorance is likely to persist within it.

These findings may be disconcerting. The first suggests that it will not do to respond to the threat of mis- and disinformation simply by advocating greater sceptical caution in incorporating the evidence available within one's social network: trustworthy information will not be efficiently assimilated if distrust runs rampant. The second finding may even suggest strategies for disinformants – such as finding ways to occupy the central node in a star network (O'Connor & Weatherall, 2019). Yet this cuts equally in the opposite direction: those wishing to ensure the dissemination of trustworthy information may equally accommodate the point – whether they be social or other (e.g. news) media organizations, or governments or charities that provide oversight. In any case, the results themselves provide a starting point upon which further research on strategies for coping with our informational environment may build, and that constrain viable approaches to combatting mis- and disinformation.

References

Ball, B., Koliousis, A., Mohanan, A., & Peace, M. (2024a). Computational philosophy: reflections on the PolyGraphs project. *Humanities and Social Sciences Communication, 11*, 186. https://doi.org/10.1057/s41599-024-02619-z

Ball, B., Koliousis, A., Mohanan, A., & Peacey, M. (2024b). Misinformation and higher-order evidence. *Humanities and Social Sciences Communications, 11*(1), 1–12

Grim, P., & Singer, D. (2022). Computational philosophy. In E. N. Zalta & U. Nodelman (Eds.), Fall 2022 edition *The stanford encyclopedia of philosophy*. Stanford University.

Kahneman, D. (2011). *Thinking, fast and slow*. Macmillan.

Mousavi, S., & Gigerenzer, G. (2017). Heuristics are tools for uncertainty. *Homo Oeconomicus, 34*, 361–379.

O'Connor, C., & Weatherall, J. O. (2018). Scientific polarization. *European Journal for Philosophy of Science, 8*, 855–875.

O'Connor, C., & Weatherall, J. O. (2019). *The misinformation age: How false beliefs spread*. Yale University Press.

Ritchie, K. (2020). Social structures and the ontology of social groups. *Philosophy and Phenomenological Research, 100*(2), 402–424.

Rosenstock, S., Bruner, J., & O'Connor, C. (2017). In epistemic networks, is less really more? *Philosophy of Science, 84*, 234–252.

Uzquiano, G. (2004). The supreme court and the supreme court justices: *A metaphysical puzzle. Noûs, 38*(1), 135–153.

Williamson, T. (2000). *Knowledge and its limits*. Oxford University Press.

Zollman, K. J. S. (2007). The communication structure of epistemic communities. *Philosophy of Science, 74*, 574–587.

War Against Fake News: Avoiding Manichaeism

Sacha Ferrari

Contents

Abstract In this article, I will analyze two strategies involved in the war against fake news: the quality rating of media outlets and fact-checking. For each of them, I will point out the motivations of the stakeholders to undertake such demanding enterprises. These motivations are allegedly for the sake of better information and democracy. In regard to the economic and sociological landscape of journalism and the inherent issues of such practices, one can doubt that these motivations are the only ones. More importantly, these strategies raise a lot of concerns: methodologically (lack of consensus among fact-checkers), consequentially (limited positive impact), ethically (self-grated authority and symbolic violence), deontologically (conflict of interest and double standard), and ontologically (naive realism). These findings undermine the Manichaean picture of savior fact-checkers fighting against vile fake news producers and give a sharper understanding of the war against fake news led by the media.

S. Ferrari (✉)
Center for Logic and Philosophy of Science, KU Leuven, Leuven, Belgium
e-mail: sacha.ferrari@kuleuven.be

M. M. Resch et al. (eds.), *Trust and Disinformation*,
https://doi.org/10.1007/978-3-031-96790-0_8

1 Introduction

Fake news was named 'World of the Year' in 2017 by Collins Dictionary.[1] Even though this epistemic phenomenon is all but new (Burkhardt, 2017), its current form still has raised concern among policy makers, academics, journalists, and human rights activists in a noticeable way. In recent years, the more prominent examples of fake news waves are to be found in the Brexit campaign (Rawlinson, 2020), the 2016 US elections (Allcott & Gentzkow, 2017), and the Covid crisis (Rocha et al., 2021). The lack of quality control of the content and the viral properties of fake news can lead to epistemic and democratic issues: echo chambers, filter bubbles, radicalization, distrust, hate speech, etc. Facing these threats, academics and other intellectuals, governments, and mainstream media are joining forces to fight this peculiar form of harmful disinformation. These fight-back strategies may take various forms: ban laws, fact-checking initiatives, media literacy educational programs, the quality rating of media outlets, (self) regulation of media, etc. Such a narrative polarizes the public debate between vile fake news producers and savior fact-checkers. In this article, I will argue that this Manichaean and dualistic picture of the war against fake news is merely a chimera and does not hold. To that end, I will first recall the socioeconomic background surrounding media today. I will then successively analyze two of the proposed fight strategies against fake news: the quality rating of media outlets and fact-checking. I will briefly describe what they consist of and show how these practices raise methodological, ethical, deontological, and ontological issues. These findings will undermine the Manichaean picture of savior fact-checkers fighting against vile fake news producers, and give a sharper understanding of the war against fake news led by the media. I will now depict the sociological landscape in which quality rating and fact-checking take place.

2 The Background

Traditional journalism has been facing a severe crisis for several years now. This crisis is complex and stems from several changes in our society. A salient manifestation of this crisis consists in the drastic drop in the number of readers, and, as a consequence, a sales drop of printed media outlets. According to the French Ministery of Culture (Ministère de la culture, 2020), 12 million printed issues were sold daily in France in 1965. This number has constantly declined to barely 5.4 million in 2020. As a consequence of this decline, journals receive less and less money from advertisers. Historically, advertisements played an important role in the financial health of newspapers. For instance, advertisements represented 60% of the revenues of the French newspaper *Le Monde* in 1970, whereas it only represented

[1] https://www.politico.eu/article/collins-dictionary-picks-fake-news-as-word-of-the-year/. Accessed 15 Mar 2023.

20% in 2010 (Poulet, 2009). Moreover, the sales price is not sufficient to cover the production cost of the newspapers. More and more media outlets are struggling to make ends meet. In France, 105 journals were available on the market, whereas there are only 60 in 2020 (Ministère de la culture, 2020). In response, European countries like France and Italy give millions of euros of state aid each year to media outlets to save them.

These financial troubles do not only stem from the lack of readers. Other reasons play a role: the development and accessibility of alternative media formats (such as television and the Internet), the progressive scarcity of press shops, the ecological incentive to digitization, the rise of fabrication cost, the rise of free journals and free content, and the decline of classified ads which constituted an important source of revenue (Poulet, 2009).

This decline in readers can also be explained by the progressive public distrust that journalism is facing. We will make the reasons explicit later in the text. Fighting against fake news has been considered by some media and journalist organizations as not only a way to regain trust from the audience (by debunking fake news and/or providing better and more transparent content) but also as a way to face digitization and generate more revenue. I want to make clear that I will not argue that journalists' war against fake news is a purely pecuniary strategy without a democratic ideal. But I will not deny either that to counteract the deep economic troubles faced by the press, new strategies are needed to gain more readers and advertisers.

In the two following sections, I will address two strategies implemented by journalists and media stakeholders. The first one consists in rating the quality of news websites and the second one consists in fact-checking other media's articles. The first one is more global and aims to give the reader a general overview of the quality of the media, whereas the second one is more fine-grained and debunks specific articles in detail. I will examine the motivations, the methodology, and the impact on trust and democracy of these two strategies.

3 Quality Rating

Facing the plethora and the diversity of media outlets, a reader can easily be lost in which media to trust or not. Some initiatives have been launched to guide readers across the variety of media outlets by giving them a quality score. For instance, *NewsGuard* was launched in 2018 to give "an online safety for readers, brands, and democracies".[2] More concretely, *NewsGard* consists in an add-on that one can install in a web browser. Each news website is given a 0–100 score based on the assessment of nine quality criteria with a detailed 'Nutrition Label'. These criteria are explicitly mentioned on their website. Among them, one can find: the website's habit of publishing false content, the regularity of error corrections, the reliability of

[2] https://www.newsguardtech.com/. Accessed 15 Mar 2023.

the website's sources, the explicit differentiation between news and opinions, the presence of disclosure of conflicts of interest, and the display of the name of the content creator. In the same vein, one can mention *Media Bias/Fact Check* (*MBFC*), an American website founded in 2015. As mentioned on their website, "the mission of MBFC is to educate the public on media bias and deceptive news practices".[3] In order to do so, they rate news websites according to two aspects: their political bias and their factual accuracy. By using a 0–10 scale, *MBFC* gives each news website a position in the political spectrum (from the extremes to the center). In addition, they investigate the factual reporting of such websites and their commitment to trustworthiness. *MBFC* praises commitment to non-partisanship, transparency of sources and funding, and a rigorous methodology. A third organization, *Transparency Map*, checks media outlets and assesses their transparency according to eight criteria: mention of the sources, attached support, updates of the situation, etc. They aim to "promote transparency in news gathering and reporting".[4]

In these three examples, we see that the assessment criteria and their respective weights in the final score differ. However, according to Lin et al. (2023) who compared the score given by six quality assessment websites (including *NewsGuard* and *MBFC*), it seems that there is "a substantial agreement" in the final score given by these websites. As the authors admit, these results have to be nuanced because not all the evaluators assess the same websites, and some of them (like *NewsGuard*) regularly update their evaluations. Furthermore, low-quality content websites are not often analyzed even if they sometimes reproduce good content coming from other high-quality content websites. However, *NewsGuard* gives a good label to journals like the *New York Times* even though it published multiple fake news articles to support the US invasion of Iraq in 2003 (Sussman, 2020).

Reducing the complexity of a media outlet to a 0–100 score seems to me slightly artificial and simplistic. I understand the desire of the evaluators to give a stable scale to compare media outlets, but I doubt that it will develop the reader's critical mind as they pretend. Media literacy is not only about receiving passive knowledge about which media to trust or not but also about acquiring the tools to assess the quality of a medium by oneself. The explanative text behind each score is scarce and oversimplifies the complexity of a media outlet's editorial board politics. I also doubt that making citizens pay $4.95 a month of subscription fees (for *NewsGuard*) will bring media literacy equally to each class of the population. Indeed, some studies show that, among other factors, income and education level are correlated to the capacity to debunk fake news (Wang, 2020, Arin et al., 2021and Pop & Ene, 2019). Those who might need it the most will not benefit from it. Lastly, concerning the efficiency of such a strategy against fake news, Aslett et al. (2022) ran a large-scale study to assess the impact of media quality labels on the readers. They found that these labels have a limited average effect on the individual's news diet quality and fail to reduce misconceptions. Another organization chose a more positive way to

[3] https://mediabiasfactcheck.com/about/. Accessed 15 Mar 2023.

[4] https://www.transparentjournalism.org/reader/. Accessed 15 Mar 2023.

assess media quality. In 2018, *Reporters Without Border* launched the *Journalist Trust Initiative* (*JTI*) project. This EU-funded project consists in a self-regulatory initiative designed to combat disinformation online (Boulay, 2018). To do so, they want to settle a white list of media outlets compliant with some ethical norms of journalism and independence: systematic editorial process to ensure the accuracy of content, sourced statistics, location reporting, identified authors, ethical treatment of violent or explicit content, diversity and training of editorial staff etc.[5] Each reliable media outlet will then receive a quality label. The scope of *JTI* is plural. First, the *JTI* label can be an incentive for other media to improve their internal process and be explicit about the methods they use. Second, this label will help "consumers, advertisers, distributors, and regulators to identify and reward trustworthy journalism".[6] Advertisers will then have a guide to better select trustworthy media and avoid those that might give them a bad reputation. Of course, this implies a gain of advertisement revenues for media compliant with *JTI* guidelines. *JTI* is transparent about the situation that journalism is undergoing. As claimed in the introductory video on their website: "Journalism today suffers from a loss of trust and revenue, from a loss of attention and respect. Its future is at stake. [...] Transparency is the only way to restore trust".

Some journalists and activists opposed initiatives such as *JTI*. According to them, this label fixes its own standard about how good journalism should be conducted, regardless of other journalistic practices that might be equally legitimate. One of the aims of the *JTI* label is indeed to serve as a form of soft auto-regulation of journalism to anticipate stronger state regulation. For instance, in France, this quality label has been incorporated in the process used for public fund allocation in journalism (Sedel et al., 2021), which raises concerns about the financial future of media outlets not *JTI* -compliant but proposing another (legitimate) way of practicing journalism. However, this anticipation has its limits. Countries like France and Germany have already implemented regulations for punishing fake news producers. As for laws against negationism, these regulations raise some issues related to the freedom of speech.

In my perspective, it is worth noticing that the position of *JTI* is not neutral and needs to be understood in a sociological field inhabiting symbolic power exchanges (Sedel et al., 2021; Bourdieu, 1977). Indeed, the field of journalism can be seen as a market of symbolic goods. These symbolic goods are reputation, prestige, respect, recognition, trustworthiness, etc. *JTI* is a self-designed actor entitled to award such symbolic goods to media outlets. When *JTI* offers a trust label to a journal, it not only certifies that this journal is compliant with some guidelines, but it also legitimizes this journal in the field of journalism by giving it some symbolic goods (and, by doing so, enhancing its symbolic capital). Journals not compliant with *JTI* guidelines will be symbolically marginalized. This marginalization can lead to what Bourdieu (1987) called symbolic violence: an implicit acceptation of a symbolic

[5] https://www.jti-app.com/report/Xlq9zOj1NoebQ1Pp. Accessed 22 Nov 2023.

[6] https://www.journalismtrustinitiative.org/about. Accessed 15 Mar 2023.

hierarchy by those who undergo it. The latter is arbitrarily imposed on them and will be tacitly reproduced by them because this norm is considered unquestionable. In light of this interpretation, I have difficulties seeing how *JTI* can create a more respectful, plural, egalitarian, and democratic mediatic ecosystem.

4 Fact-Checking

In this article, I will consider news fact-checking as the activity of verifying the truthfulness of published pieces of news. News fact-checking can be done by a journal (for instance, *Les Décodeurs* of the French newspaper *Le Monde*, *Reality Check* of the *BBC*, *Reality Check* of *The Guardian* and *Fact Checker* of the *Washington Post*) or by an organization (for instance, *Factcheck.org* which is run by the *The Annenberg Public Policy Center* at the University of Pennsylvania). Other fact-checkers focus instead on politicians' statements such as *PolitiFact* launched by the *Tampa Bay Times* to evaluate Obama's claim during his 2008 campaign (Amazeen, 2015). The latter assessment process is called political fact-checking. The frontier between news fact-checking and political fact-checking is sometimes narrow, especially when politicians cite journal articles to strengthen their points. Fact-checking can be seen as a more precise strategy to fight fake news than quality rating. There are nowadays hundreds of fact-checking agencies. Some of them joined the *International Fact-Checking Network* (*IFCN*). Launched by *Poynter* in 2015, this association aims at harmonizing fact-checking agencies by the settlement of a code of ethics. *Mutatis muntandis*, it is the equivalent of the *JTI* for fact-checkers.

In my perspective, the main issue is the epistemic premises that fact-checking supposes. Believing a claim is not only influenced by the credibility or trustworthiness of the piece of information but also by the epistemic and social ecosystem surrounding the reader (Alloing & Vanderbiest, 2018; Margolin et al., 2018): their political views, the availability of other sources of information, their social class, etc. Namely, the epistemic proximity with the source, the preexistent beliefs, the psychological bias (confirmation bias, rebuttal strategies, etc.) and political affiliations often overpass the appeal to credibility or coherence. Correcting a false belief is a complex process that cannot always be solved by displaying the correct piece of information to the reader. As an example, political affiliation seems to play an important role in belief updating. Democrats and Republicans will react differently to fact-checking (Walter et al., 2020). Democrats are not only those who are more prone to correcting their beliefs (Walter et al., 2020), but are as well those who are more prone to consuming fact-checking websites (Robertson et al., 2020). As we will see later in this section, fact-checkers demonstrate a more negative attitude toward conservative media outlets (Bozarth & Budak, 2020). Thus, it is not surprising that fact-checking websites please more to a Democrat audience than to a Republican one.

These epistemic misconceptions explain the mitigated efficacy of fact-checking in correcting false beliefs. Some literature reviews analyzed articles treating the impact of fact-checking. According to Nieminen & Rapeli (2019), some studies "find that fact-checking reduces misperceptions, others that corrections are often ineffective". Moreover, these studies highly focus on US media and its audience. In their meta-analysis, Walter et al. (2020) point out the limited impact of fact-checking: "Though fact-checking can be used to strengthen preexisting convictions, its credentials as a method to correct misinformation (i.e., counter-attitudinal fact-checking) are significantly limited". This can be explained partially by the fact that individuals are more prone to adopting pro-attitudinal correction instead of correcting their beliefs in a way that can contradict their prior political views. This effect is even more stringent during electoral campaigns when citizens are even more overwhelmed by fake news than usual. However, another meta-analysis based on a different set of articles (Chan et al., 2017) points out the strong effect of misinformation correction. The former authors (Walter et al., 2020) explain this discrepancy by the fact that they focus exclusively on political misinformation. This type of misinformation includes more resistance and partisanship than other types of misinformation (e.g., about health or marketing) considered in Chan et al.'s article. That is a possible explanation for why Walter et al.'s meta-analysis points toward a more limited impact of fact-checking than Chan et al. Facing such a lack of *statu quo* in the literature regarding fact-checking efficiency, it seems to me hard to justify not only the fact-checking relevance for correcting readers' false beliefs caused by fake news consumption but also the amount of money and effort put into it if it were only for educational purposes. Indeed, as we have seen, fact-checkers cannot correct political misinformation efficiently in the context of an electoral campaign, which is originally one of their main goals. As mentioned in Section "The Background", there is far more at stake for journalists in this fight against fake news: regain of trust, respectability, esteem, readers, advertisers, and financial health.

In addition to its approximate efficacy, fact-checking undergoes some methodological issues caused by implicit biases. As Uscinski & Butler (2013) note, fact-checkers only analyze the information they deem relevant to check. It includes focusing on specific topics, countries, politicians, or media. This bias is inherent to fact-checking given that a comprehensive check of the thousands of news articles published each day is virtually impossible. Actually, this pragmatic discrimination undermines the so-called objectivity of fact-checkers and makes salient some of their biases. Indeed, Marietta et al. (2015) point out that the three fact-checkers they analyzed (*FactCheck.com*, *PolitiFact* and *The Washington Post*) do not consider each of the three considered subjects in this study (racism, climate change, and public debt) as equally important (in term of coverage). But a stronger bias is also at stake. For instance, statements claiming the reality of climate change were not checked. Only those against the reality of climate change were. However, one cannot deny that some positive articles about climate change could turn out to be fake news (for instance, an article artificially inflating the value of temperature increase in the coming decades to stress out the catastrophic impact of climate change).

Fact-checkers in this study consider not only climate change as real *prima facie*, but also never consider positive news about climate change as questionable. Even if obvious claims such as the 'the Earth is round' and '2 + 2 = 4' are virtually uncontested, there are still correct ways and incorrect ways to justify them. And fact-checkers should be concerned about it. In my view, spreading exaggerated news (and, hence, fake news) about the impact of climate change will not improve trust. Even if climate change acceptance or denial does not only rely on sole epistemic arguments but also on prior political views, these exaggerated news could even be counterproductive and could undermine any genuine and honest dialogue. In contrast with climate change, fact-checkers diverge on other topics. For instance, on the consequences of the growth of public debt, *Politifact* only checked articles claiming a negative effect of the public debt, whereas the two other fact-checkers addressed the positive and negative statements the same way. Stated differently, Marietta et al. points out that even when fact-checkers consider the same topic, they evaluate positive and negative articles in a different way. These results indicate that fact-checkers assume a set of unquestionable realities and are more reluctant to fact-check information confirming their prior beliefs. This exhibits a lack of objectivity when approaching some topics. Bozarth & Budak (2020) confirm the existence of some biases of fact-checkers: the latter demonstrates a more unfavorable attitude against small and conservative-leaning credible news sites. Articles from these sources were more often labeled incorrectly as fake news than those from more liberal sources.

Another methodological problem lies in the lack of consistency between different fact-checking organizations. If fact-checking is an objective and rigorous practice, we would expect the conclusion concerning an article to be the same regardless of the fact-checker. Lim (2018) compared the outcomes of two fact-checkers: *Politifact* and *Fact Checker*. The first issue pertains to the fact that both fact-checkers only share 10% of their analyzed statements in common. Furthermore, these statements were rarely exactly the same, which makes a comprehensive study difficult. Although fact-checkers always agree on extreme news (i.e., obvious truth or obvious hoax), their conclusions diverge on milder pieces of news (i.e., statements more subject to interpretation or ambiguity). What is the usefulness of fact-checking then? Evident truths or lies are easily debunkable and can be done with a few clicks on the internet by any layperson. It is especially when things get more complex or blurry, that the help of fact-checkers is potentially really useful. However, the latter do not agree on the outcomes, which makes fact-checking an illusory objective and universal practice. I do not mean that fact-checkers are not rigorous, I just mean that a lot of statements in the description of realities are not black or white and an unequivocal and unproblematic description is an almost impossible task. As mentioned in the last paragraph, even trivial scientific truths can be justified by ambiguous explanations. Of course, this equivocity is even more salient when talking about political and social matters. For instance, consider a news article that claims that "the burglaries increased in the neighborhood by 20% after the migrants came into the city" (Bernecker, 2021, p. 289). If we assume that the statistic is really correct (i.e., that the police monitored 20% more burglaries and the rise of foreigners is proven by the population registry), what is the status of such a piece of information? Of course, the way it is framed might suggest that the migrants are responsible for burglaries, which might not be true. Is that fake news fabricated for political aims? Or a simple conservative editorial bias? Or an objective – yet ambiguous – statement pointing to a correlation and not causation between two events? There is leeway for interpretation and I doubt that, even with the most sincere commitment to rigor and objectivity, all fact-checkers will consider it the same way.

This commitment to rigor and objectivity seems also jeopardized by the conflicts of interest and double standards practiced by some fact-checkers. Indeed, many media both produce their own news and have their own team of fact-checkers (*Le Monde*, *The Washington Post*, *The Guardian*, etc.). So they are both the judge and the jury. One could have hoped for a more objective and independent position for fact-checkers. Ironically, some of these media outlets produced fake news in the past. One can remember the statements of Julian Assange published by *The Guardian* which turned out to be a pure creation of the author (Greenwald, 2016), and the recurrent misinformation about the Russian government by *The Washington Post* (which is also the recipient of the Pulitzer Prize and listed as trustworthy by *IFCN*) (Greenwald, 2017). My point is not to determine if there was a clear intent to deceive from these two journals or if they are just unintended errors (but I doubt this) or editorial bias. Merely, I want to point out that, for some fact-checking journals, the idiom "it's the pot calling the kettle black" seems a good description of the double standard of some fact-checkers. I do not mean that there is no self-quality control assessment in these media outlets. Some newspapers have implemented peer-reviewing by other colleagues before publication and have internal procedures to correct or retract erroneous articles (e.g. after some reader complains to an ombud).

Lastly, the practices of fact-checking and quality rating of media outlets have an issue in common. Like the *JTI*, the *IFCN* is a self-granted agent determining what good fact-checking practice is and what it is not. Note that *IFCN* included *Politifact* in their white list. One can question the conflict of interest in this choice given that *Politifact* is funded by the *Tampa Bay Times*, which is itself a branch of *Poynter*. One can also question the neutrality of *IFNC* when it is awarded a 13.2 million dollars grant from Google[7] and when it allies with Facebook to fight the spread of Covid-related fake news.[8] In addition to what precedes, these practices raise even more questions about the neutrality and independence of fact-checker organizations.

---

[7] https://blog.google/outreach-initiatives/google-news-initiative/how-google-and-youtube-are-investing-in-fact-checking/. Accessed 15 Mar 2023.

[8] https://www.facebook.com/journalismproject/coronavirus-grants-fact-checking. Accessed 15 Mar 2023.

5 Discussion and Conclusion

In the preceding discussion, I analyzed two strategies involved in the war against fake news: the quality rating of media outlets and fact-checking. For each of them, I pointed out the motivations of the stakeholders to undertake such a time and money-consuming enterprise. These motivations are allegedly for the sake of better information and democracy. In regard to the economic and sociological landscape of journalism and the inherent issues of such practices, one can doubt that these motivations are the only ones. To stress this point, I showed that there was no consensus in the literature about the efficiency of both quality rating and fact-checking. More importantly, these strategies raise a lot of issues and concerns: methodological (lack of consensus among fact-checkers), ethical (self-granted authority and symbolic violence), deontological (conflicts of interest, double standards), and ontological (presupposition of some elements of reality, naive realism).

In light of my conclusions, one can question if such fights against fake news are the best way to regain trust in the media and improve democracy. It is important to remember that the lack of trust in the media is correlated not only to the rise of the Internet but also to the change classic journalism has been undergoing for 50 years. Until the 1970s, many popular media were highly partisan and targeted very specific audiences. In 1974, the Watergate scandal made the headlines thanks to the restless work of two journalists of *The Washington Post*. This case study was progressively considered as the new standard for good journalism and investigative journalism became, for many, the new trend (Poulet, 2009). As a result, many journals in the US and Europe left their partisan identity to endorse a more objective and neutral way of doing journalism. By an irony of fate, it is at this precise moment that the press started disconnecting from the readers and lost many of them. One of the reasons lies in the fact that most of the traditional media while praising objectivity and denying any partisanship, took political positions in major recent elections. For instance, in Europe, traditional media outlets such as *Le Monde* or *The Guardian* were more positive toward the adoption of the Maastricht treatise during the 2005 French referendum, toward Hilary Clinton during the 2016 presidential US elections, and toward the maintenance of the UK in the EU during the Brexit referendum. The outcomes of these elections have been against all odds. Some journalists saw these results not only as a democratic failure but also as a failure of journalists who did not achieve to gain the trust of the readers (Michel, 2017, Lefilli^atre, 2016, Lordon, 2016; Poulet 2009).

The case of journalism is just an example of more global distrust from the citizens towards institutions in general. Ambiguous relations of media with politics and economy (for instance, the billionaire Jeff Bezos owns *The Washington Post*) put journals in a difficult position in terms of trust. In particular, the collaborations of media rating and fact-checking organizations with Facebook, Google, or governments do not improve trust. This distrust is exacerbated each time one of these organizations claims the monopoly of objectivity and implicitly patronizes the reader about news to trust or not. This form of symbolic power is exactly one of the main reasons for the ongoing distrust. Using it again in the war against fake news will just not work or might even be counterproductive.

Maybe a better alternative in the fight against fake news is to accept the complexity of the social reality and the illusive commitment to neutrality. I do not expect *journals* to be impartial, but I expect the *press* to be. Like in history, it is by comparing the narratives of different media that one can have a more lucid description of an event. Diversifying one's media diet seems to me a valid autonomous and efficient way of reaching a simulacrum of objectivity (if any). Of course, it is asking citizens a lot to daily consume many different kinds of news. It is more probable that the most privileged of them could really achieve it. In addition, such an aggregation of sources will lead to a more reliable version of the news only if the diversity of perspectives is guaranteed. This plurality could be jeopardized if the press is ruled by governments or by a small group of shareholders. Accepting the plurality of the journals, their different perspectives, and refusing arbitrary epistemic authority seem to me in line with the democratic idea of public debate. And does not this show that inter-rater variability of fact-checking is a sign of a healthy ecosystem?

Maybe fake news is an urgent threat to our democracy and must be fought (Jaster & Lanius, 2021). Or maybe fake news is not a problem at all and the war against it produces more harm than good (Coady, 2021). Or maybe fake news is just not a real epistemic phenomenon and we should stop using this term (Habgood-Coote, 2019). In this article, I did not take a position in this triad. In my opinion, neither fake news nor a fight against it is problematic *per se*. The problem is to polarize the public debate between vile fake news producers and savior fact-checkers. In light of the previous discussion, this Manichaeism is merely a chimera and does not hold. Fact-checkers and media quality assessors are not only driven by democratic purposes but also by the willingness to regain trust and financial stability. Furthermore, the *ethos* of these two practices demonstrate the use of symbolic power, methodological weaknesses, conflicts of interest, double standards, and disputable ontological simplifications. For these reasons, I assert that defending such an ingenious binary vision of the war against fake news is equivalent to extracting two epistemic phenomena (fake news and the war against them) out of their intelligible environment. Fake news start being considered as a problem nowadays because some social conditions are met (e.g., public distrust and financial troubles of the media). Paradoxically enough, in echo with the Hegelian master-slave dialectics, the media involved in this fight became somehow dependent on fake news (and the war against them) for their financial future and for reestablishing their epistemic authority in a climate of global distrust.

Acknowledgement Part of this work was supported by the Research Foundation Flanders (Fonds Wetenschappelijk Onderzoek, FWO), Grant No. G0B8616N. The author thanks Sylvia Wenmackers for her insightful review of this article's daft.

References

Allcott, H., & Gentzkow, M. (2017). Social media and fake news in the 2016 election. *The Journal of Economic Perspectives, 31*, 211–236.

Alloing, C., & Vanderbiest, N. (2018). La fabrique des rumeurs numériques. Comment la fausse information circule sur Twitter? *Le Temps des médias, 1*, 105–123.

Amazeen, M. (2015). Revisiting the epistemology of fact-checking. *Critical Review, 27*, 1–22.

Arin, K. P., Lacomba, J. A., Lagos, F., Mazrekaj, D., & Thum, M. P. (2021). Misperceptions and fake news during the Covid-19 pandemic. In *CESifo working paper*. https://doi.org/10.2139/ssrn.3842330

Aslett, K., Guess, A. M., Bonneau, R., Nagler, J., & Tucker, J. A. (2022). News credibility labels have limited average effects on news diet quality and fail to reduce misperceptions. *Science Advances, 8*, eabl3844.

Bernecker, S. (2021). An epistemic defense of news abstinence. In S. Bernecker, A. K. Flowerree, & T. Grundmann (Eds.), *The epistemology of fake news* (pp. 286–309). Oxford University Press.

Boulay, Eric. 2018, RSF and its partners unveil the journalism trust initiative to combat disinformation. Retrieved March 20, 2023, from https://rsf.org/en/rsf-and-its-partners-unveil-journalism-trust-initiative-combat-disinformation

Bourdieu, P. (1977). La production de la croyance. *Actes de la recherche en sciences sociales, 13*, 3–43.

Bourdieu, P. (1987). *Distinction: A social critique of the judgement of taste*. Harvard University Press.

Bozarth, L., & Budak, C. (2020). Toward a better performance evaluation framework for fake news classification. *Proceedings of the International AAAI Conference on Web and Social Media, 14*, 60–71. https://doi.org/10.1609/icwsm.v14i1.7279

Burkhardt, J. M. (2017). History of fake news. *Library Technology Reports, 53*, 5–9.

Chan, M.-P. S., Jones, C. R., Jamieson, K. H., & Albarracín, D. (2017). Debunking: A meta-analysis of the psychological efficacy of messages countering misinformation. *Psychological Science, 28*, 1531–1546.

Coady, D. (2021). The fake news about fake news. In S. Bernecker, A. K. Flowerree, & T. Grundmann (Eds.), *The epistemology of fake news* (pp. 68–81). Oxford University Press.

Greenwald, G. (2016). The Guardian's summary of Julian Assange's interview went viral and was completely false. In *The intercept*.

Greenwald, G. (2017). Washpost is richly rewarded for false news about Russia threat while public is deceived. In *The intercept*.

Habgood-Coote, J. (2019). Stop talking about fake news! *Inquiry, 62*, 1033–1065.

Jaster, R., & Lanius, D. (2021). Speaking of fake news. In S. Bernecker, M. K. Flowerree, & T. Grundmann (Eds.), *The epistemology of fake news*. Oxford University Press.

Lefilli^atre, J. (2016). En France, les médias promettent de "réduire la distance avec les lecteurs". In *Libération*.

Lim, C. (2018). Checking how fact-checkers check. *Research & Politics, 5*, 205316801878684. https://doi.org/10.1177/2053168018786848

Lin, H., Lasser, J., Lewandowsky, S., Cole, R., Gully, A., Rand, D. G., & Pennycook, G. (2023). High level of correspondence across different news domain quality rating sets. *PNAS Nexus, 2*, pgad286. https://doi.org/10.1093/pnasnexus/pgad286

Lordon, Frédéric. 2016. Politique post-vérité ou journalisme post-politique? Le Monde diplomatique

Margolin, D. B., Hannak, A., & Weber, I. (2018). Political fact-checking on twitter: When do corrections have an effect? *Political Communication, 35*, 196–219.

Marietta, M., Barker, D. C., & Bowser, T. (2015). Fact-checking polarized politics: Does the fact-check industry provide consistent guidance on disputed realities? *The Forum, 13*, 577–596.

Michel, Pascal. 2017. Post-vérité et fake news: fausses clartés et points aveugles. Retrieved March 10, 2023, from https://www.acrimed.org/Post-verite-et-fake-news-fausses-clartes-et

Ministère de la culture. 2020. Le tirage de quotidiens de 1945 à 2020. Retrieved March 10, 2023, from https://www.culture.gouv.fr/Thematiques/Presse/Documentation/Chiffres-Statistiques

Nieminen, S., & Rapeli, L. (2019). Fighting misperceptions and doubting journalists' objectivity: A review of fact-checking literature. *Political Studies Review, 17*, 296–309.

Pop, M. I., & Ene, I. (2019). Influence of the educational level on the spreading of fake news regarding the energy field in the online environment. *Proceedings of the International Conference on Business Excellence, 13*, 1108–1117.

Poulet, B. (2009). *La fin des journaux et l'avenir de l'information*. Editions Gallimard.

Rawlinson, F. (2020). *How press propaganda paved the way to brexit*. Palgrave Macmillan.

Robertson, C. T., Rachel, R., & Mour˜ao, and Emily Thorson. (2020). Who uses fact-checking sites? the impact of demographics, political antecedents, and media use on fact-checking site awareness, attitudes, and behavior. *The International Journal of Press/Politics, 25*(2), 217–237.

Rocha, Y. M., de Moura, G. A., Desidério, G. A., de Oliveira, C. H., Lourenço, F. D., & de Figueiredo Nicolete, L. D. (2021). The impact of fake news on social media and its influence on health during the COVID-19 pandemic: A systematic review. *Journal of Public Health, 31*, 1007–1016. https://doi.org/10.1007/s10389-021-01658-z

Sedel, J., Ouakrat, A., Pacouret, J., & Nou^s, C. (2021). Présentation du dossier. un journalisme de qualité ? Hiérarchisations et classements des actualités. *Political Communication, 16*(1), 5–12. https://doi.org/10.3917/pdc.016.0005

Sussman, G. (2020). Making enemies: the mainstream media spectacle and US foreign policy. *Perspectives on Global Development and Technology, 19*(1–2), 138–156.

Uscinski, J. E., & Butler, R. W. (2013). The epistemology of fact checking. *Critical Review, 25*(2), 162–180. https://doi.org/10.1080/08913811.2013.843872

Walter, N., Joshua Cohen, R., Holbert, L., & Morag, Y. (2020). Fact-checking: A meta-analysis of what works and for whom. *Political Communication, 37*(3), 350–375.

Wang, T.-L. (2020). Does fake news matter to election outcomes? The case study of Taiwan's 2018 local elections. *Asian Journal for Public Opinion Research, 8*(2), 67–104.

Targets of Countermeasures against Disinformation: Individual Beliefs and Group Beliefs

Lei Niu

Contents

Abstract Disinformation has been observed to affect individual beliefs, leading to the adoption of false beliefs or the reduced likelihood of accessing and evaluating true information. Accordingly, previous research has largely focused on individual beliefs-oriented countermeasures to combat disinformation. However, the impact of disinformation on group beliefs has been overlooked in these discussion. Through a review of relevant cases, it becomes clear that the influence of disinformation on groups cannot be solely attributed to individual beliefs. Consequently, individual beliefs-oriented strategies are not sufficient, and a comprehensive approach to countering disinformation should target both individual beliefs and group beliefs. In light of this, a group beliefs-oriented strategy is proposed as a potential means to effectively combat disinformation.

L. Niu (✉)
University of Cologne, Cologne, Germany

M. M. Resch et al. (eds.), *Trust and Disinformation*,
https://doi.org/10.1007/978-3-031-96790-0_9

1 Introduction

The direct epistemic result of disinformation is that people tend to question their true beliefs and they are more likely to hold false beliefs. More importantly, disinformation can have impacts not only on individuals but also on groups. On the direct interpretation, the more individuals are influenced by disinformation, the more the group is affected. Many philosophers have recently argued that groups as a whole can function as epistemic agents and generate beliefs that can be differentiated from individual beliefs that are held by individuals and common beliefs that are shared by all or most members. This raises important questions about whether disinformation can influence group beliefs and whether countermeasures targeting group beliefs should differ from those targeting individual beliefs.

Two main kinds of strategies against disinformation have been proposed in the literature.[1] The first is disinformation-directed strategies that aim to eliminate disinformation through the regulations of governments and media companies. In particular, AI technology can be used to detect and eliminate disinformation before consumers receive it. When this is successful, there is no need to worry about the influence of disinformation. However, this strategy faces several challenges.

First of all, there is a definitional problem with disinformation.[2] Key factors, including intentions, accuracies of information and corresponding harms, are usually difficult to identify and measure. Policies based on uncertainty and weighing these factors differently can exclude and legitimate some harmful disinformation and prohibit some true information. Even extensive and strict censorship may not work, because it can sometimes reinforce the distrust of epistemic authorities and it faces ethical charges, such as conflicting with the human right of free speech.[3] In addition, lacking international cooperation among countries with different political values and laws also adds difficulties, so sophisticated detection and elimination are often not applicable.

For the sake of insufficiency of eliminating disinformation before people receive it, the second strategy is information consumers-directed, which aims to improve individuals' epistemic capacities. Various methods can be used to prevent people from being influenced by disinformation, such as adding authentic media programs, learning from experts, improving critical thinking, and having face to face conversations. Although this strategy is deemed necessary, its effectiveness may be limited, as disinformation can have a long-lasting influence even after it has been eradicated and true information has been presented. Considering the mechanisms of the proliferation of disinformation, such as new communication technologies and new platforms, there can be a discrepancy between what people believe and what people share.[4] Since disinformation is sometimes more attractive to be reposted and shared,

[1] As for therapies and countermeasures, see, for instance, the final section in Bernecker et al. (2021).

[2] See, for instance, Lange & Lechterman (2021).

[3] See, for instance, Gläßel & Paula (2020).

[4] In previous work, Sarah Wright (2021) have explored such discrepancy.

it is challenging to mitigate its influence. Apart from disinformation, many factors can exacerbate the information crisis, such as cognitive biases and social environments, given that disinformation can become intertwined with conspiracy theories and echo chambers.[5] Accordingly, presenting only accurate information or refuting disinformation may be inadequate since individuals may still falsely draw inferences based on their available information.

Both strategies aimed at mitigating the impact of disinformation face challenges and require further adaptation. It is important to note that individual beliefs may not be the entire target, but the primary goal of these strategies is to prevent individuals from consuming disinformation and to correct people's questionable individual beliefs. In this sense, I classify these strategies as individual beliefs-oriented strategies. In this paper, I will consider the potential influence of disinformation on group beliefs. The immediate impact can be that a group suspends judgment so that it does not form a belief, and holds a questionable belief, regardless of whether it believes a false proposition or disbelieves a true proposition.

In this paper, I will illustrate how disinformation can potentially impact group beliefs and argue that false group beliefs can be different from the mere aggregation of individual beliefs. In other words, a group belief can be irreducible to individual beliefs, so that group beliefs-oriented strategies should also be included in order to fight against disinformation. The plan of this paper is as follows: In section "Disinformation and Group Beliefs", I will define paradigmatic cases of false group belief and highlight some cases where group beliefs diverge from individual beliefs, and argue that the correctness of group beliefs cannot be entirely through the correctness of individual beliefs. In section "Group Beliefs-Oriented Strategy", I will motivate group beliefs-oriented strategies to combat consequential epistemic phenomena. I conclude in section "Conclusion".

2 Disinformation and Group Beliefs

The matter of justified group belief is a subject of dispute among epistemologists.[6] While a complete elucidation of this issue is beyond the scope of this paper, there exist paradigmatic instances wherein groups hold false beliefs or fail to formulate beliefs altogether. When it comes to questionable group beliefs, one might immediately consider the summative account of group beliefs as the foremost conceptual framework[7]:

Summative account of group belief: A group has a belief that p if and only if all or most (operative) members of the group believe that p.

---

[5] See, for instance, Nguyen (2020).

[6] See, for instance, Backes (2021), List (2005), List & Pettit (2011), Silva (2019).

[7] For more discussions of the summative account of group beliefs, see Gilbert (2004) and Lackey (2020).

According to this view, when a group is deceived, it means that all or most people, or only operative members, are deceived and hold questionable beliefs. Group beliefs, in this sense, are reducible to individual beliefs, and a successful attempt to change group beliefs means changing the beliefs of all or most members. This is the common case and the individual beliefs-oriented strategy is accordingly sufficient. However, there are cases where it is feasible for a group to hold a belief that is not held by all or most of its individual members. If so, the summative account is subject to serious counterexamples, such as the following examples:

1. All philosophers jointly develop an argument for skepticism and conclude that we do not have a hand, while they believe otherwise as non-philosophers.
2. A court believes that a person is innocent, while all jurors individually believe otherwise because of hearsay.
3. A community refuses to get vaccinations because disinformation worsens the trust relationship between experts and laypeople.
4. A research group jointly conceals the truth and lies to the public because a large amount of money is paid.

These group beliefs that are not reduced to individual beliefs are dubbed the non-summative account.[8] In cases (1) and (2), the group in question is an epistemic agent who aims at attaining truth, and the formation of questionable group beliefs stems from epistemic reasons. Conversely, non-epistemic reasons play a vital role in cases (3) and (4), and groups fail to be desirable epistemic agents. Given the disparity between individual beliefs and group beliefs, it is possible for there to be cases in which disinformation solely impacts group beliefs rather than individual beliefs. In other words, a group holding false or questionable beliefs is not identical with at least some individuals holding false or questionable beliefs, and a group failing to form beliefs is not identical with at least some individuals failing to form beliefs.

The extent of the disparity between individual beliefs and group beliefs depends on the precise nature of a group epistemic agent. In order to capture the intuition that groups can have false beliefs and to explain the examples I listed above, I take the joint acceptance account of group belief to be paradigmatic. This is because it represents the most commonly accepted form of a non-summative account of group belief, and can easily explain the ontology of groups, wherein the decision-making procedure necessitates joint acceptance. Additionally, several features that differentiate group beliefs from individual beliefs can be provided by the joint acceptance account of group beliefs and the neglected impact of disinformation can be exemplified. The following definition can be considered in this regard[9]:

> Joint acceptance account: A group believes that p if and only if all or most members, or only operative members, jointly accept that p.

---

[8] See, for instance, Lackey (2020).

[9] Proponents of joint acceptance account include, for instance, Gilbert (1987), Tuomela (1992), Mathiesen (2006), and de Ridder (2022).

In contrast to individual beliefs and the summative account of group beliefs, it is crucial to note that at least two conditions of the non-summative account must be fulfilled:

1. The group evidence condition, which necessitates the formation of a group belief based on group evidence or total evidence. For one thing, individuals can vote for a proposition they do not believe in. For another, a jointly accepted belief can be a composite incorporating the beliefs of different members when each member has different epistemic tasks. Thus, a group belief can be supported by aggregating all members' evidence or voted evidence, and a group belief in this account can be different from individual beliefs that are held by individuals and common beliefs that are shared by all or most members of a group.
2. The group's epistemic responsibility condition, which requires groups to satisfy joint acceptance and to be epistemic agents who pursue the truth and avoid errors. Believing a proposition can be different from accepting a proposition, and it is then possible that individuals in a group refuse to jointly commit to a belief that they actually hold. In this sense, joint acceptance can be an epistemic requirement for forming beliefs. In other words, although acceptance is commonly a state that is under voluntary control, group acceptance can be involuntarily sensitive to epistemic goals because of some decision procedures. Correspondingly, both conditions (1) and (2) can be potentially jeopardized by disinformation. In what follows, I will illustrate some situations in which disinformation affects group beliefs rather than individual beliefs in relation to these two conditions.

2.1 *Disinformation and Group Evidence*

The first condition of the joint acceptance account reveals that group beliefs can be divergent from individual beliefs in terms of evidence. In order to promote the wisdom of crowds, it is always expected that groups are able to consider and unify all relevant evidence that all members actually possess and should possess. If this is correct, then there are possible cases where disinformation affects only group evidence rather than individual evidence, which undermines the condition (1). To illustrate this situation in a bit more detail, let us consider the following example:

> The emperor's new clothes: Two deceivers offer an emperor imaginary clothes and claim that the clothes are invisible to people who are stupid. Everyone, including the emperor, sees nothing, but people jointly pretend to believe otherwise. In the end, a child yells the truth, and everyone starts to trust and express their own beliefs. However, the procession continues.[10]

[10] As for collective ignorance, see also Bjerring et al. (2014) and de Haan (2021).

The group belief in this story satisfies both (1) the evidence condition because the story about the clothes spread throughout the country and no one publicly questioned it, and (2) the epistemic responsibility condition because of joint acceptance. However, the group forms a belief solely on the basis of disinformation, which all members of the group jointly accept. Thus, it is reasonable to conclude that the group is deceived and holds a false group belief, even though it may not necessarily follow that all individuals are independently deceived. The divergence of deception between group beliefs and individual beliefs in this case is twofold. Firstly, the group as a whole rather than any of its members fails to exclude disinformation. Based on sensory evidence, everyone can easily know that the emperor is naked, so that mere disinformation cannot defeat one's belief and no one has false individual beliefs. However, the key evidence every member possesses is entirely excluded by the group when it makes decisions.

One may object that the epistemic mistake illustrated by this example can still be classified as a false individual belief, because individuals mistakenly interpret others' beliefs and believe that other people are smart enough to see the clothes. People who can see the clothes are thus their epistemic superiors. The belief of others functions as higher-order evidence and defeats one's own evidence. In this sense, it can be argued that all members have a false individual belief, which consequently results in a false group belief. However, it seems that people do not have a decisive epistemic reason for deferring to others, particularly if one discovers that their epistemic peers and inferiors maintain that the clothes are magnificent. As such, individuals have the option of either suspending judgment or maintaining their own beliefs, rather than automatically deferring to the belief of others. In other words, the false group belief can be reduced to a false individual belief, but it is not necessarily so.

Secondly, the group rather than its members fails to update evidence in time. This failure to update occurs after the child yelled the truth about clothes. Once new evidence about the beliefs of others is made known, all members are able to identify the disinformation and confirm that the emperor wears nothing. However, because there is no valid and timely procedure in place for reviewing new evidence for a group, the latter remains deceived and continues to proceed with the procession for the nonexistent clothes. In this regard, groups are sometimes more vulnerable than individuals in terms of disinformation.

The community in the case above can be understood as an unorganized group that is not formed purposefully. Not only unorganized groups, but also organized groups can be influenced by disinformation. Consider the following example:

> To secure professional promotion, some researchers fabricate pictures, statistics, and peer review reports. Some intentional fabrications can be verified only by substantive experiments that go beyond the mechanisms of the review process. In light of this, science journals and research foundations are deceived.

Generally, each editor and referee is accountable for a different portion of these review works, and they collectively accept papers and research proposals. While it is reasonable to state that groups, i.e., science journals and research foundations, are deceived, its operative members are not necessarily deceived. For one thing, their task or main task is not to verify all the data[11] provided by submitters, and some fabricated data, such as stealing others' ideas or including the names of non-contributors, in practice cannot be verified in a short time. The roles and duties of editors and referees can vary across different fields.[12] Nevertheless, it can be granted that at least the verification of some data is not their tasks. For another, their vote for individual acceptance is not necessarily equivalent to their own individual beliefs about the overall quality of these data. In other words, groups and its operative members can have different responsibilities, assigning different epistemic weights to the same evidence or taking different evidence into consideration. Therefore, disinformation is admissible for groups but not for individuals.

In both cases, the disparity between the evidence available to groups and individuals explains why the former is more susceptible to disinformation. When disinformation targets group evidence, the correctness of individual beliefs may not be necessary nor sufficient for the correction of erroneous group beliefs. Put differently, when individuals have true beliefs or the admitting of group evidence exceeds the epistemic duty of individuals, individual belief-oriented strategy is not sufficient. In order to prevent groups from being influenced by disinformation, strategies increasing the ability of groups to detect information and consider reliable evidence are needed.

2.2 *Disinformation and Group Epistemic Responsibility*

Disinformation not only impacts group evidence, but also group epistemic responsibility. This means that even when groups and individuals have access to the same evidence, groups as a whole can be more vulnerable to disinformation than individuals. In regard to the joint acceptance account of group beliefs, group epistemic responsibility entails at least two requirements. Firstly, a decision must be jointly accepted by all or most members of the group. Secondly, groups as a whole should be epistemic agents who pursue epistemic goals. Disinformation can hinder groups from fulfilling both of these requirements, thereby compromising the group epistemic responsibility. In this section, I provide two arguments for identifying the potential connection between disinformation and group epistemic responsibility.

The first argument involves distinct epistemic goals regarding groups and its members. Put simply, there is an epistemic goal of avoiding errors, which can be distinguished from the goal of attaining truth.[13] When presented with the same evidence, if one hopes to avoid believing what is false, this epistemic goal can be achieved by suspending judgment until there is overwhelming evidence in favor of

[11] By data, I generally refer to all kinds of material used in a scientific study.

[12] For discussions about different peer review procedures, see for instance Horbach & Halffman (2019).

[13] For a discussion of different epistemic goals, see for instance Kelly (2013) and Mathiesen (2011).

a proposition. In other words, one can reasonably disbelieve what their evidence supports. However, suspending judgment is not sufficient when the goal is to attain the truth, and one should believe what their evidence fits. The idea is that group agents and their members can have different epistemic goals and then weigh evidence differently. Diverse epistemic goals can potentially undermine group epistemic responsibility, such as joint acceptance. In the context of disinformation, because of the decision-making requirements of groups, it is possible that groups as a whole are more susceptible to disinformation than individuals, resulting in undesirable decisions. Consider the following example:

> A virus spreads widely throughout the community. There is a vaccine that is effective against the virus, but it requires that all or most members jointly accept vaccination. Disinformation about variants of the virus, side effects and others' negative attitudes towards vaccines proliferates on social media. Although side effects have been repeatedly disproved by scientific evidence, epistemic free-riders who temporarily suspend judgment and refuse to accept vaccination can possibly enjoy the benefit of others' joint acceptance. By noticing the existence of epistemic free-riders, everyone refuses to accept vaccination.[14]

In this case, individuals and groups have different epistemic goals as a result of conflicts between self-interests and group-interests. Everyone has the motivation to access the truth, but avoiding what is false is more vital for individuals, especially since people have become aware of the existing epistemic free-riders. When the majority of the community is motivated to avoid believing what is false and to engage in free-riding behavior, those who accept vaccination cannot contribute to the collective effort to stop the spread of the virus, and they certainly bypass what is true and possibly believe what is false since they bear the potential risks of side effects. In contrast, when the majority jointly accepts vaccination, people can freely ride on the efforts of others without bearing the possible side effects. Avoiding believing what is false for individuals is epistemically reasonable until there is sufficient evidence, and sometimes a cautious attitude could be an epistemic virtue. However, when every individual suspends judgment, groups as a whole lack the motivation to access the truth and ultimately believe what is false. This can be epistemically worse, for the reason that groups fail to undertake their epistemic responsibility.

Disinformation can pollute the epistemic environment and worsen the trust relationship among members of groups, and exacerbate the epistemic disparity between individuals and groups, creating barriers to the formation of group beliefs. Distrust can make individuals value the group being right less than their own being right. Consequently, being an epistemic free-rider is the dominant strategy that can maximize both their epistemic and non-epistemic benefits in the group. Given that groups and its members have different epistemic goals, individual beliefs-oriented strategies that aim to reinforce reliable evidence are insufficient for reconciling these

---

[14] As for vaccination, see also John (2011) and Goldenberg (2016).

conflicting epistemic goals and incentivizing individuals to act in the epistemic interest of groups.

The case above illustrates a situation where epistemic considerations can impede the fulfilment of the group epistemic responsibility condition (2). My second argument involves non-epistemic considerations. In addition to forming problematic group beliefs, a more profound problem is that groups will never form any beliefs at all. In other words, non-epistemic factors, such as political motivation, self-interests and different values, can undermine the requirement of condition (2), leading to groups falling short of their role as epistemic agents committed to the pursuit of truth.

The most noteworthy non-epistemic factor are pragmatic considerations. There are many cases where epistemic considerations are parallel to pragmatic considerations. Put differently, accessing the truth and avoiding errors can maximize the best interests of groups. However, scenarios do exist in which the use of false or incorrect information can contribute to non-epistemic benefits. In this sense, increasing polarization can contribute to the spread of disinformation, and disinformation can conversely exacerbate the imbalance between epistemic considerations and pragmatic considerations. Ultimately, people's individual beliefs deviated from what they jointly accepted as a member of their group. Consider the following example:

> Different candidates prepare for a presidential election. Disinformation about these candidates proliferates on social media. Most disinformation cannot be debunked timely and thus people naturally suspend judgment. However, in order to dismiss their opponents, some political groups accept and repost the disinformation.

It is plausible that the group possesses no more reliable evidence than the individuals. However, groups can jointly accept obvious false propositions that no member believes. Suspending judgment until sufficient evidence is available constitutes a natural response in the face of disinformation. Nevertheless, groups typically exhibit a greater inclination to make decisions based on unverified information or disinformation. What supports the group to make the judgment and supports the group to lie is not their evidence that is sensitive to truth, but pragmatic considerations. In this regard, group beliefs are no longer subjective to epistemic evaluations. It might be debatable whether group beliefs based on political motivation or practical considerations are a bad thing, but there are clear cases in which group beliefs based on epistemic considerations are expected.

Various measures can be used to bolster the individuals' motivation for accuracy. Financial incentives, for example, have demonstrated effectiveness in encouraging information consumers to discern disinformation.[15] However, it is not reasonable to state that all members lack the ability as well as the motivation to access the truth in cases of questionable group beliefs, since knowing the truth is helpful in designing sophisticated strategies for groups concealing the truth. In this sense, approaches that try to correct individual beliefs by means of deleting disinformation and educating information consumers cannot mitigate the epistemic detrimental effects.

[15] For discussions, see, for instance, Rathje et al. (2023).

As we can see, both organized and unorganized groups can form beliefs that are distinct from individual beliefs. Although groups may differ in their structures and functions, the joint acceptance account of group belief exemplifies some situations in which disinformation has unsymmetrical impacts on group beliefs and individual beliefs. In addition, concerns about evidence, epistemic goals, and pragmatic considerations can highlight the divergence between individual beliefs and group beliefs. One important lesson that can be drawn from these divergences is that correcting individual beliefs may not always be necessary nor sufficient for addressing group beliefs. In the next section, I consider what it means to have a group beliefs-oriented strategy against disinformation.

3 Group Beliefs-Oriented Strategy

According to the individual beliefs-oriented strategy, the fewer disinformation people meet and the more epistemic capabilities people have, the less detrimental effects disinformation has. It is expected that this approach can prevent deception and promote epistemic benefits for groups by merely collecting individual epistemic improvements. The efficiency and feasibility of this strategy against disinformation entirely depend on how we can maximize the epistemic improvements for as many people as possible. However, as illustrated, there are cases where groups are deceived while its members are not, so the epistemic improvement of group beliefs cannot be solely attributed to the accuracy of individual beliefs. Moreover, it is sometimes even impossible for groups to have beliefs, because of the failure of joint acceptance and epistemic responsibility. Thus, a group beliefs-oriented strategy is needed in its own right, with the aim of improving the epistemic abilities of groups as a whole to function epistemically better than individuals.

Although the specific measures that can be taken to improve the epistemic abilities of groups will depend on the structures, functions, and aims of groups, the nature of group beliefs themselves, as well as concrete disinformation, two kinds of general considerations must be included regarding the features of group beliefs. First, in addition to fact-checking and disinformation-checking, epistemic interferences, such as presenting more information and restricting some information, towards groups should be considered. For the reason that groups and individuals can have different evidence and disinformation has impacts only on groups' evidence rather than individuals' evidence, barriers of improving the epistemic abilities of individuals and groups can be distinct. One kind of epistemic intervention is to initiate restrictions on group evidence. Some group evidence can be denied, diminished and discouraged. For instance, juries are not permitted to consider hearsay evidence, even if individuals may not voluntarily choose to disregard it. This illustrates how groups can disregard certain types of information as a collective. It is then expected that groups can disregard disinformation as a whole when they form group beliefs, even if the members may not necessarily agree with the group's position by suspending judgments or rejecting the group beliefs. In addition to collecting reliable group evidence, group epistemic abilities can be enhanced by regulating group decision procedures and facilitating the process of interpreting and updating information, which can make groups rely more on verified evidence and be more likely to form true beliefs.

Second, non-epistemic interferences, such as coercion and legal restrictions, play an important role in improving the epistemic benefits of groups. As I illustrated above, it is important to acknowledge that group agents may fail to undertake epistemic responsibility even when they possess sufficient evidence and epistemic abilities. Group epistemic agents and their members can naturally have divergent epistemic goals, which can result in a discrepancy between what is true on an individual level and what is true on a group level. In light of this, epistemic interventions are not sufficient, and a sophisticated non-epistemic intervention should be designed to motivate groups and its members to take on group epistemic responsibilities. Furthermore, some group agents are more sensitive to non-epistemic considerations than individuals, which can impede their ability to function as epistemic agents. As a result, groups can jointly accept propositions based on disinformation and spread disinformation. In this sense, non-epistemic interferences, such as increasing the cost of avoiding the truth and accepting false beliefs, should be considered.

The target that is interfered with is the group itself and its members. It is necessary to specify the types of practical considerations that should be restricted or discouraged to achieve group epistemic goals. It is commonly agreed that other-regarding harmful choices and beliefs should be prohibited. In addition, some beliefs that are not clearly other-regarding harmful and that undermine the justification and the formation of group beliefs, can be discouraged. For example, some unverified information can potentially challenge the trust relationship between governments and citizens, or between experts and laypeople. Moreover, conflicting pragmatic considerations between different individuals as well as different groups can undermine the formation of group beliefs, leading to unstable and varying beliefs even when no epistemic condition has changed. Such conflicts can also result in deception and the spread of disinformation among groups. In order to fight against disinformation, interferences with groups that aim to identify potential conflicts and manage conflicts of interest is thus needed. Therefore, non-epistemic countermeasures should strive to align practical considerations with the epistemic goals of believing the truth and avoiding errors.

4 Conclusion

The paper has sought to provide a preliminary analysis of the link between disinformation and group beliefs. I have argued that disinformation can have an influence on both unorganized groups, such as communities, and organized groups, such as governments and science journals. Because of disinformation, groups can believe a false proposition or fail to believe a true proposition. In addition, disinformation can prevent groups from acting as epistemic agents who pursue the truth and avoid errors. Moreover, I have considered that three features, i.e., group evidence, group epistemic goals, and non-epistemic considerations, can distinguish false group beliefs from false individual beliefs. Due to these divergences, group epistemic agents may hold false beliefs that none of its individual members possess, and groups may lack beliefs that all or most members possess. Correspondingly, the barriers to correcting and changing group beliefs and individual beliefs can be distinct. This provides us with stronger reasons to focus on false group beliefs and design group beliefs-oriented strategies against disinformation. Considering the features of group beliefs, two general considerations should be included, namely, epistemic interferences towards groups on the one hand, and non-epistemic interferences on the other.

The analysis has focused primarily on the joint acceptance account of group beliefs. However, the mechanism and the cause of disinformation are complex, and groups always differ in structure and function, so that only one kind of intervention cannot solely prevent individuals and groups from being negatively impacted by the pernicious effects of disinformation, and not all interventions can meet the technical and ethical challenges. Despite these limitations, the new category can deepen our perspective on the influence of disinformation, and new countermeasures need to be devised.

References

Backes, M. (2021). Can groups be genuine believers? The argument from interpretationism. *Synthese, 199*(3–4), 10311–10329.

Bernecker, S., Flowerree, A. K., & Grundmann, T. (Eds.). (2021). *The epistemology of fake news*. Oxford University Press.

Bjerring, J. C., Hansen, J. U., & Pedersen, N. J. L. L. (2014). On the rationality of pluralistic ignorance. *Synthese, 191*(11), 2445–2470.

de Haan, N. (2021). Collective culpable ignorance. *Thought: A Journal of Philosophy, 10*(2), 99–108.

de Ridder, J. (2022). Group belief reconceived. *Synthese, 200*(2), 1–21.

Gilbert, M. (1987). Modelling collective belief. *Synthese, 73*, 185–204.

Gilbert, M. (2004). Collective epistemology. *Episteme, 1*(2), 95–107.

Gläßel, C., & Paula, K. (2020). Sometimes less is more: censorship, news falsification, and disapproval in 1989 East Germany. *American Journal of Political Science, 64*, 682–698. https://doi.org/10.1111/ajps.12501

Goldenberg, M. J. (2016). Public misunderstanding of science? Reframing the problem of vaccine hesitancy. *Perspectives on Science, 24*(5), 552–581.

Horbach, S., & Halffman, W. (2019). The ability of different peer review procedures to flag problematic publications. *Scientometrics, 118*, 339–373.

John, S. (2011). Expert testimony and epistemological free-riding: The MMR controversy. *The Philosophical Quarterly, 61*(244), 496–517.

Kelly, T. (2013). Evidence can be permissive. In M. Steup & J. Turri (Eds.), *Contemporary debates in epistemology* (pp. 298–311). Blackwell.

Lackey, J. (2020). *The epistemology of groups*. Oxford University Press.

Lange, B., & Lechterman, T. (2021). Combating disinformation with AI: Epistemic and ethical challenges. In *IEEE international symposium on ethics in engineering, science and technology (ETHICS)* (Vol. 1, pp. 1–5).

List, C. (2005). Group knowledge and group rationality: A judgment aggregation perspective. *Episteme, 2*(1), 25–38.

List, C., & Pettit, P. (2011). *Group agency: The possibility, design, and status of corporate agents*. Oxford University Press.

Mathiesen, K. (2006). The epistemic features of group belief. *Episteme, 2*(3), 161–175.

Mathiesen, K. (2011). Can groups be epistemic agents? In H. B. Schmid, D. Sirtes, & M. Weber (Eds.), *Collective epistemology* (pp. 20–33). Ontos.

Nguyen, C. T. (2020). Echo chambers and epistemic bubbles. *Episteme, 17*(2), 141–161.

Rathje, S., Roozenbeek, J., & Van Bavel, J. (2023). Accuracy and social motivations shape judgements of (mis)information. *Nature Human Behaviour, 7*, 892–903. https://doi.org/10.1038/s41562-023-01540-w

Silva, P. (2019). Justified group belief is evidentially responsible group belief. *Episteme, 16*(3), 262–281.

Tuomela, R. (1992). Group beliefs. *Synthese, 91*, 285–318.

Wright, S. (2021). The virtue of epistemic trustworthiness and re-posting on social media. In S. Bernecker, A. K. Flowerree, & T. Grundmann (Eds.), *The epistemology of fake news* (pp. 245–264). Oxford University Press.